Lecture Notes in Earth Sciences

Vol. 1: Sedimentary and Evolutionary Cycles. Edited by U. Bayer and A. Seilacher. VI, 465 pages. 1985.

Vol. 2: U. Bayer, Pattern Recognition Problems in Geology and Paleontology. VII, 229 pages. 1985.

Vol. 3: Th. Aigner, Storm Depositional Systems. VIII, 174 pages. 1985.

Vol. 4: Aspects of Fluvial Sedimentation in the Lower Triassic Buntsandstein of Europe. Edited by D. Mader. VIII, 626 pages. 1985.

Vol. 5: Paleogeothermics. Edited by G. Buntebarth and L. Stegena. II, 234 pages. 1986.

Vol. 6: W. Ricken, Diagenetic Bedding. X, 210 pages. 1986.

Vol. 7: Mathematical and Numerical Techniques in Physical Geodesy, Edited by H. Sünkel. IX, 548 pages. 1986.

Vol. 8: Global Bio-Events. Edited by O.H. Walliser. IX, 442 pages. 1986.

Vol. 9: G. Gerdes, W.E. Krumbein, Biolaminated Deposits. IX, 183 pages. 1987.

Vol. 10: T.M. Peryt (Ed.), The Zechstein Facies in Europe. V, 272 pages. 1987.

Vol. 11: L. Landner (Ed.), Speciation of Metals in Water, Sediment and Soil Systems. Proceedings, 1986. VII, 190 pages. 1987.

Lecture Notes in Earth Sciences

Edited by Somdev Bhattacharji, Gerald M. Friedman,
Horst J. Neugebauer and Adolf Seilacher

11

Lars Landner (Ed.)

Speciation of Metals in Water, Sediment and Soil Systems

Proceedings of an International Workshop,
Sunne, October 15–16, 1986

Springer-Verlag
Berlin Heidelberg GmbH

Editor

Dr. Lars Landner
Swedish Environmental Research Group
Götgatan 35, S-11621 Stockholm, Sweden

ISBN 978-3-540-18071-5   ISBN 978-3-540-47733-4 (eBook)
DOI 10.1007/978-3-540-47733-4

2132/3140-543210

PREFACE

It is to-day generally recognized by environmental scientists that the particular behaviour of trace metals in the environment is determined by their specific
physico-chemical forms rather than by their total concentration. With the introduction, several years ago, of atomic absorption spectrometry at many laboratories
involved in environmental studies, a technique for simple, rapid and cheap determination of total metal concentrations in environmental samples became available.
As a consequence, there is a plethora of scientific papers and reports where metal
concentrations in the environment are only reported as total concentrations.

It appears that the simplicity of making accurate determinations of total metal
contents in water, sediment and biological samples has somewhat masked the need
for improved knowledge about the various forms of metals occurring in the environment as well as the bioavailability of these forms. In other words, the need for
metal speciation in studies of metals in the environment does not seem to have
become obvious to most environmental scientists until relatively recently. As
a matter of fact, it was only in the middle of the 1970s that the first systematic
attempts were made to obtain information about the various metal species occurring
in environmental samples.

During the last ten years, however, a revolutionary change of attitude towards
the importance of metal speciation has occurred and considerable research effort
has been devoted by environmental scientists to measuring the concentrations of
biologically important trace metals in surface waters. There is currently an
increasing effort to couple the development of chemical analytical techniques
to process-related biological problems. Concurrently, a new focus is being imposed
on ecological impact studies, that of determining which active trace metal species
merit the most intensive research from the standpoint of environmental perturbation. Current efforts are directed towards the development of chemical speciation
schemes which can be related directly to measures of bioavailability (1).

This considerable growth of interest in the field of metal speciation has
resulted, during the last few years, in the organisation of a number of workshops
and seminars dealing partly or exclusively with chemical speciation. The first
one in this new category of international scientific meetings seems to have been
the NATO Workshop on "Trace Element Speciation in Surface Waters and Its Ecological
Implications", held in Nervi, Italy, in November 1981 (2). Later, the Dahlem

Conferences, Berlin, took up the same topic when a workshop on "The Importance of Chemical Speciation in Environmental Processes" was organised in September 1984 (3). It might be pertinent to mention in this context also the International Seminar on "Speciation, Separation and Recovery of Metals", held in 1986 (4).

When the present workshop on "The Speciation of Metals in Water, Sediment and Soil Systems" was organised in Sunne, Sweden, it was recognised that research in this field requires a multidisciplinary team approach. Therefore, scientists with many different backgrounds – engineers, physicists, hydrologists, geologists, analytical chemists, biologists and ecologists – were invited with the aim of forcing them to find a common language and, as far as possible, a mutual under-standing. The purpose of the workshop was to review and evaluate the recent scien-tific progress in the field of metal speciation and related subjects and to discuss, from the various perspectives of the participants, priorities for planning of future research and possible areas of common interest and cooperation. The programme of the workshop included critical reviews of the latest progress in the development of analytical methods for separation and determination of various metal species in water, sediment and soil, a discussion of why and when metal speciation is to be recommended, a review of the major environmental factors influencing the distribution and transformation of metal species, and a discussion of the implica-tions of this approach to metal research for environmental planning and management.

The Sunne Workshop, which was attended by about 30 scientists from several different countries, started with presentations of five major background papers, followed by a number of contributed papers. The integral text of all these contri-butions is contained in this volume. During the second day of the workshop, the participants gathered in four separate Working Groups to discuss various scenarios wherein speciation of metals would be useful or necessary. The development of adequate analytical techniques, both for routine work and for in-depth scientific studies, was discussed and future needs were considered. A particularly fruitful aspect of these discussions was the close contact achieved between the analytical chemists, who develop and run the analytical procedures, and the biologists, who make use of the analytical data so as to interpret the bioavailability and effects of metals in the environment. It became quite clear that a much closer cooperation between the two sides is necessary – and possible – to further our knowledge on the distribution and effects of metals in the environment.

REFERENCES

1. Leppard, G.G. Trace element speciation and the quality of surface waters: An introduction to the scope for research. In: ref. 2: 1–10.

2. Leppard, G.G., Ed. Trace Element Speciation in Surface Waters and Its Ecolo-gical Implications. Proc. NATO Advanced Research Workshop, Nov. 2–4, 1981, Nervi, Italy. (New York: Plenum Press, 1983).

3. Bernhard, M., F.E. Brinckman, and P.S. Sadler, Eds. The Importance of Chemical Speciation in Environmental Processes. Dahlem Conferences, September 2-7, 1984, Berlin. (Berlin: Springer Verlag, 1986).

4. Patterson, J.W., Ed. Speciation, Separation and Recovery of Metals. Proc. of an International Seminar. (Chelsea: Lewis Publ., 1986).

TABLE OF CONTENTS

INTRODUCTION

The Metal Conference in Athens, 1985: A Growing Interest in Metal
  Speciation; A Review
  Rudolf Reuther 3

SECTION 1: ANALYTICAL TECHNIQUES FOR SPECIATION OF METALS
           DETECTION AND ROLE OF MOBILE METAL SPECIES

Metal Speciation in Solid Wastes - Factors Affecting Mobility
  Ulrich Förstner 13
Analytical Techniques in Speciation Studies
  Brit Salbu 43
Approaches to Metal Speciation Analysis in Natural Waters
  G.M.P. Morrison 55
Metal Fractionation by Dialysis - Problems and Possibilities
  Hans Borg 75
Trace Element Speciation in Natural Waters Using Hollow-Fiber
  Ultrafiltration
  E. Lydersen, H.E. Bjørnstad, B. Salbu and A.C. Pappas 85
The Importance of Sorption Phenomena in Relation to Trace Element
  Speciation and Mobility
  B. Allard, K. Håkansson and S. Karlsson 99

SECTION 2: BIOLOGICAL IMPLICATIONS OF METAL SPECIATION

Testing the Bioavalaibility of Metals in Natural Waters
  Peter Pärt 115
Case Studies on Metal Distribution and Uptake in Biota
  Olle Grahn and Lars Håkanson 127
Effects of pH on the Uptake of Copper and Cadmium by Tubificid
  Worms (Oligochaeta) in Two Different Types of Sediment
  Anders Broberg and Gunilla Lindgren 145
Aluminium Impact on Freshwater Invertebrates at Low pH: A Review
  Jan Herrmann 157

SECTION 3

Summary of Working Group Reports
  Lars Landner 179

APPENDIX

The Workshop Participants 189

Introduction

THE METAL CONFERENCE IN ATHENS, 1985 : A GROWING INTEREST IN METAL SPECIATION

A REVIEW

Rudolf Reuther
Swedish Environmental Research Group
Fryksta, S-665 00 KIL

The 5th International Conference on Heavy Metals in the Environment, which took place in Athens (Greece), in September 1985, was called, in advance, one of the biggest scientific meetings in the field of environmental research held in Europe. Looking at the great number of authors (1112) and papers presented (433) from all over the world, the conference framework was indeed impressive. Its real importance for our future understanding of the environmental impact of heavy metals, as a basis for an effective health and environmental protection policy, has still to be evaluated. The aim of the present review is not to assess the importance of the conference, but just to examine to what extent the papers presented were dealing with metal speciation. Furthermore, it will discuss the general approaches to metal speciation used in the middle of the 1980s, as well as the precise methods selected for separation and detection of various metal species.

Almost one fourth of all papers published in the Conference Proceedings (2 volumes, 1337 pp) dealt with speciation of metals. Some of the papers were directed to the determination of at least one analytically defined chemical form (e g tetraalkyllead), others used a more general approach, where groups of species were separated (e g low and high molecular weight compounds) (1).

However, the use of the term "species" still seems to cause some confusion: generally it was used in a descriptive manner and not as a measurable quantity related to e g the "reactivity" of the metal in the environment. So far we may rely upon the definition suggested by the participants of the Dahlem Workshop on "The Importance of Chemical Speciation in Environmental Processes", where a "species" is understood as the "molecular representation of a specific form of an element" (2).

An overview of the frequencies (x) of environmental samples and metals studied, as well as separation and detection methods applied in the work accounted for in the Conference proceedings, is presented in Table 1.

Table 1. The frequency (x) of environmental matrices, metals, separation and
detection methods used in speciation studies.

Environmentel Matrix	Metals	Separation Methods	Detection Analysis
o biological specimens (24x)	o Cu (37x)	o mechanical (filtration, (9x) centrifugation, seiving)	A A S (39x)
o freshwater (15x)	o Pb (34x)	o solvent extraction (8x)	A S V (7x)
o waste solids (14x)	o Cd (32x)	o ion-exchange and gel (13x) chromatography	X R D (5x)
o marine sediments (13x)	o Zn (32x)		I C P (5x)
o soil substrata (10x)	o Fe (20x)	o gas chromatography (10x)	radio tracer (3x)
o atmospheric constituents(9x)	o Ni (13x)	o HPLC Chromatography (3x)	N A A (3x)
o marine water (4x)	o Cr (11x)	o chemical extraction (27x)	I S E (3x)
o freshwater sediments (4x)	o Hg (11x)	o Catechol violet, EDTA and batho phenanthro-	M S (3x)
o waste water (1x)	o Al (8x)	line complexation,	TEM/SEM (2x)
	o Co (7x)	chemical modelling,	E S R (2x)
	o Sn (5x)	density and magnetic	XAFS (2x)
	o As (4x)	separation, sorption	o Mössbauer spectroscopy, NMR,
	o Ca, Mg, Ba,	experiments	PDC, EDX; DPP, CSV, EDA, FPM,
	K, V, U, Tl,		photospectroscopy, thermogra-
	Si, Ti, Be		vity, magnetometry, derivative
			spectroscopy

Most concern was dedicated to biological samples (e g cells, tissues, organs,
whole organisms: 24 x), freshwater samples (15 x), solid wastes (sewage sludge,
mine tailings, dredged material: 14 x), marine-estuarine sediments (13 x), soil
substrates (10 x) and airborne constituents (e g dust, fly ash, aerosols, vapour,
atmospheric water: 9 x). However, no speciation studies among those presented in
the proceedings were dealing with colloidal systems.

Speciated metals

The heavy metals copper (37 x), lead (34 x), cadmium (32 x) and zinc (32 x)
were most frequently subject to speciation procedures, both in laboratory and model-
ling studies ($\underline{1}$:475, $\underline{2}$:394[*]) as well as in routine environmental monitoring ($\underline{1}$:380,
$\underline{1}$:585, $\underline{2}$:34, $\underline{2}$:443, $\underline{2}$:454, $\underline{2}$:559). The great interest in these metals was mostly
related to their toxic character and their wide-spread occurrence in many natural
systems as a result of releases from a multitude of human activities (e g combustion, ore smelting, metal plating).

Speciation studies with iron (20 x) and manganese (14 x), mainly with respect
to their oxidic-hydroxidic forms, have been undertaken because of their well-known
implications in soil weathering and sedimentary processes. These metals might act
both as mobile scavengers for trace metals in solution and as an energy source
for certain bacteria (e g Thiobacillus ferroxidans, see $\underline{1}$:246, $\underline{1}$:266).

The speciation of aluminium (8 x) was generally attributed to its well estab-

---

* see Conference Proceedings, vol.: page

lished toxic effects (as "labile" Al) in acidified soils and freshwaters (forest disease, fish kills; see 1:700, 2:378, 2:443, 2:446).

Among other heavy metals, the following have attracted relatively great interest in speciation studies: nickel (13 x: 2:98), mercury (11 x), chromium (11 x: 2:350, 2:427), cobalt (7 x: 2:381), tin (5 x: 2:385, 2:466, 2:531, 2:537) and arsenic (4 x: 2:7, 2:10, 2:484). Some of these form organic derivatives, which may be more crucial and/or reactive in their biogeochemical behaviour (volatile, lipid-soluble, persistent) than the corresponding inorganic forms (e g methyl mercury: 1:88, 1:103, 1:282, 2:59, 2:295, 2:513, methylarsenicals, arsenobetaine, organotin compounds). There was only one paper dealing with chromium (III) and chromium (VI) species (see 2:460). No investigation has taken up the speciation of heavy metals like bismuth and molybdenum or of metalloids such as selenium, antimony, germanium and tellurium. Earth and alkali metals, like calcium, magnesium, barium and potassium (1:263, 2:334) played only a minor role in speciation studies. The same is true for vanadium (2:478, 2:522), uranium (1:113), thallium (2:286), silicium, titanium (2:375) and beryllium (2:454).

Separation methods

The most frequently used separation techniques for differentiation of the bulk concentration of metals into physico-chemically defined fractions were:
1. Chemical extraction (27 x; see Table 2);
2. Ion-exchange/gel chromatography (13 x; see Table 3);
3. Filtration (1:113), centrifugation and sieving (1:174, 2:394, 2:519) procedures (9 x);
4. Selective solvent extraction (8 x: 1:171, 1:628, 2:537).

Sorption (2:394, 2:519), and complexation experiments with synthetic solutions (e g EDTA treatment: 1:475, 2:516, or the Catechol Violet determination of Al: 1:700), separation with regard to density (2:375) and numeric chemical modelling (i e calculating stability and rate constants: 2:225, 2:295) have been applied only marginally.

While chemical extraction of metals bound to sediment, soil and solid waste, both as single leaching steps and as modified sequential processing, was routinely employed in order to obtain "operationally defined metal phases" (3) or "solid phases which contain metal species" (2), the use of gel filtration and permeation chromatography offered a means for separating organic, metal-binding fractions of different molecular weight (e g amino acids, polypeptids, proteins, humic acids) from the labile, i e the free ionic or weakly complexed, metal species (see Tables 2 and 3).

The latter separation method was also used in combination with gas chromatography (10 x: 1:82, 1:153, 1:171) and high performance liquid chromatography (3 x: 1:313,

Table 2. Chemical extraction techniques used for metal speciation.

extractants	metal species	kind of sample	vol./page
o 1 M NH₄Ac, pH 4.8	sorbed/loosely bound metals	Fe-rich concretions	I / 263
		spring water	/ 266
1 M NH₂OH·HCl	Mn oxides	(Fe, Mn, Zn, Cs, Mg,	
1 M NH₂OH·HCl + 4 M HAc	hydrous Fe oxides	K, Si)	
0.1 M Na₄P₂O₇, pH 11	organically bound metals		
6 M HCl	well ordered Fe oxides		
o 0.1 M EDTA	metal-humic acid compl.	urban sludge	I / 336
0.01 M HCl	metals bound to humics	(Mn, Zn, Cu)	
0.1 M HCl	" " " "		
1 M HCl	metal oxides + humic complexes		
Na-hydroxide tetraborate, pH 9.6	organo-mineral complexes	soil matter	I / 342
Na-pyrophosphate, pH 9.7	" "	(Cd, Cu)	
oxalate buffer, pH 3.0	amorphous/crystallized Fe and Al oxyhydroxides		
o NH₂OH·HCl, acid	reducible fraction	marine sediments	I / 348
H₂O₂, acid	oxidizable fraction	(Fe, Cr, Zn)	
o 25% HAc	non-detrital phase	est. susp. part.	I / 380
HF/aqua regia	detrital phase	matter	
		(Zn, Cu, Pb, Cd)	
o HAc	exchangeable metals	estuar. sediment	I / 384
NH₂OH·HCl	Mn oxides	(Mn, Cu, Zn, Cd, Fe,	
NH₄-oxalate buff.	Fe oxides	Pb)	
H₂O₂	organic bound metals		
acid digest.	residual fraction		
o EDTA	removal of phosphoryl-, carboxyl-, sulfhydryl- and hydroxyl-groups of membrane protein	cell wall struct. components in sewage sludge (Zn, Pb, Cr)	I / 475
o 0.1 M CaCl₂	correlation between bio-availability and plant uptake	sewage sludge-soil (Zn, Cu, Ni)	I / 484
DTPA			
o H₂O	water soluble	estuar.salt marsh	I / 545
NH₄Ac	exchangeable	sediment	
dithionite-citrate-bi-carbonate buffer	red. Fe phases	(Fe, Mn,Zn, Cu, Cd, Pb)	
HCl/HNO₃	insoluble fraction		
o 1M KNO₃	exchangeable metals	sewage sludge	I / 582
0.5 M KF	sorbed metals	(Cd, Cu, Zn, Ni, Pb)	
0.1 M Na₄P₂O₇	org. bound metals		
0.1 M Na₂EDTA	carbonate bound metals		
6 M HNO₃	sulfides		
o HAc	leachability/immobili-zation of Zn, Cu and Pb	synthetic wastes silicate sources of cement, silicate fly ash	I / 585
o as in I / 384	Tl	soils	II/ 286
o 1 M NH₄Ac, pH 7	exchangeable metals	soil samples	II/334
acid digestion	total metal conc.	(Mn, Cu, Fe, Pb, Ni, Zn, Cd, Co, Ba, Mg, Ca)	
o 1 M MgCl₂, pH 7	sorbed phase	sewage sludge	II/350
NH₂OH·HCl	metallic coatings	(Cu, Cr, Zn)	
H₂O₂/HNO₃	organic bound phase		
HF/HClO₄	crystalline phase		
o centrifugation	pore water dissolved phase	marine sediment	II/353
1 M NH₄Ac	exchangeable metals	(Cd, Ag)	
o 0.5 N HCl	easily extractable phase (= bioavailable)	coastal susp.m. and sediment	II/363

Table 3. Ion-exchange and gel chromatography techniques used in speciation studies.

ion-exchnge/chelsting resin	metsl species	kind of sample	vol./pege
Chelsx	ion-exchsngesble Mn	riverine water	I / 246
Sephadex G-75	low m.w. Pb-binding proteins	erythrocytes	I / 313
G-15 gel filtration chromat.	Pb species	lumen of the small intestine of rats	I / 423
Sephadex G-75	metallothionein (MT) like Cu complexes, un-bound Cu and Cu bound to excluded protein	"Mimulus guttatus"	II / 47
OAE Sephsdex A-25 (strong anion-exchanger)	MT enriched Cu fractions	grass "Deschampsia cespitosa"	II / 52
Bio-gel P-60 and Bio-gel P-30 (sequence of gel chromatography)	Cu linked to low m.w. proteins, high m.w. Cd binding proteins	"Euglena" (phytoflagellate)	II / 55
Sephadex G-75	metalbinding MT-like proteins of Hg, Zn, Cd	larvae of "Mytilus galloprovincialis"	II / 58
Sephadex G-75	in combinstion with. Ni-66 in protein fractions	in liver/kidney of rats	II / 98
Bio-gel P-60	low and high m.w.fract-ions of Cu and Cd bind. proteins	freshwster pulmonats "Physa acuts"	II /171
Sephacryl S-300 HPLC chromatography	high m.w. fractions low m.w. proteins (< 20,000 D)		
Sephadex G-75 combined with S-300 and S-400 gel filtr. and dialysis processes	Cd-binding high m.w. com-pounds	freshwater bivalve "Uni elongatulus Pf."	II/177
Chelex-100 and ASV	weak soluble Cu-com-plexes and electrochem. labile Cu	marine bay water	II /249
Dowex 50W-X8 (cation-exchange resin)	non-labile Al	surface freshwater	II /446

<u>2</u>:55) attached to an atomic absorption or mass spectrometry detector, in order to further isolate and quantitatively purify metal-binding organic compounds.

For both chemical extraction and gel chromatography, there seems to be a distinct need for development of more advanced specification or refining methods. This might include improved purification of high molecular weight metal-binding proteins and non-metallothionein compounds by means of further developed gel chromatographic sequences (2:55, 2:171), studies of the correlation between metal accumulation in organisms and sorption (2:397), or combined application of spectroscopic and X-ray devices (1:266, 1:585, 2:375, 2:466: infra-red-, nuclear magnetic resonance-, ultra violet-, mass spectrometry and X-ray diffractometry). Such approaches would provide additional information about structural and binding properties of the chromatographed and/or extracted organic and inorganic metal fractions. However, the low recovery of metal species from gel chromatographic methods precludes this separation mode for quantitative speciation.

There was no study using metal fractionation by in situ dialysis. On the other hand, radioactive tracers have been used for toxicological assays of metal species (^{115m}Cd in 2:110; ^{203}Pb in 1:424) and for sorption experiments (^{48}V in 2:522).

Detection methods

To determine the metal concentration in the fractionated samples, atomic absorption spectrometry (AAS), both in the flame, graphite furnace and hydride generation mode, has proved to be effective due to its operational ease, high sensitivity and the low detection limits ($10^{-10} - 10^{-12}$ g) for most metals, even with a complex sample matrix (39 x).

Electrochemical methods, such as anodic stripping voltammetry (10 x: 1:122, 1:384, 2:438), ion selective electrodes (3 x: 2:283, 2:372) or catodic stripping voltammetry (1 x: 2:481), measuring the free and/or weakly complexed metal ions, have been used to a lesser extent. These have a high analytical sensitivity and low detection limits ($10^{-6} - 10^{-10}$ g), but are supposed to manipulate the species distribution in the sample solution during the deposition and stripping step.

There was no metal determination reported using potentiometric stripping analysis (PSA).

One paper dealt with the correlation between the magnetic properties of fine solid particles and metal species, i e the correlation of adsorbed metals with the magnetic susceptibility of small particles (see 2:363).

Preliminary conclusions

Although we are still lacking a detailed and quantified knowledge of what is really happening when metal species become subjected to our fractionating and analytical "intervention", we may assume that they will adhere to the energetically most opportune reaction paths, as decisively anticipated by our experimental equipment and the designed methodology. Sampling, storage, preparation, separation and detection modes used can modify and consequently manipulate the original species of a metal and the species composition in the sample, e g by changing the species determining parameters (such as pH, temperature, light intensity, etc).

Nevertheless, the Metal Conference in Athens showed some good promises to overcome these operationally implied difficulties by the use of conserving sample treatment (sample handling in a glove-box with an inert atmosphere, immediate speciation after pretreatments) and by combining already existing methods which are sensitive, non-destructive and applicable in situ/in vivo (e g ion selective electrodes, nuclear magnetic resonance (NMR), electron spectroscopy for chemical analysis (ESCA) or Mössbauer spectroscopy: 1:384, 1:423, 2:375, 2:466).

Another aspect, related to the impact of metals on biota (essential, indifferent or toxic metal species), was almost completely neglected in the speciation studies presented at the Conference. This was the combination of epidemiological studies with monitoring of metal species, i e investigations of the occurrence of a certain metal species along with the detection of significant health anomalies or biological injuries.

A more routine involvement of "reference material" (see the National Bureau of Standards, U.S.) and standardized speciation schemes (4, 5), e g "metal speciation handbooks", would help to intercalibrate and compare data from different study areas and environmental sites. An improved comparability between different studies would also facilitate a rapid implementation of effective remedial measures against metal pollution.

In conclusion, there is a clear need to upgrade the Water and Soil Quality Standards for trace metals to a level where our present knowledge on chemical speciation of metals is taken into account, because the species distribution is crucial for our understanding of the biogeochemical cycling, the fate and the toxicity of metals.

References

1. International Conference on Heavy Metals in the Environment. Athens - September 1985. Vol. 1:751 pp; Vol. 2:586 pp. CEP Consultants Ltd.

2. Bernhard, M., F.E. Brinckman, and P.S. Sadler, Eds. The Importance of Chemical Speciation in Environmental Processes. Dahlem Conferences, September 2-7, 1984, Berlin. (Berlin: Springer Verlag, 1986).

3. Tessier, A. and P.G.C. Campbell. Partitioning of Trace Metals in Sediments: Relationships with Bioavailability. Presented at the Internat. Workshop on "In-situ Sediment Contaminants", Aberystwyth, Wales, U.K., 28 pp. (1984).

4. Hart, B.T. and S.H.R. Davies. Estuarine coastal Mar. Sci. 12: 353-374 (1981).

5. Salomons, W. and U. Förstner. Environ. Technol. Lett. 1: 506-517 (1980).

Section 1

Analytical Techniques for Speciation of Metals
Detection and Role of Mobile Metal Species

METAL SPECIATION IN SOLID WASTES - FACTORS AFFECTING MOBILITY

Ulrich Förstner
Arbeitsbereich Umweltschutztechnik,
Technische Universität Hamburg-Harburg
Postfach 90 14 03, D-2100 Hamburg 90

Abstract

The availability of trace metals for metabolic processes is closely related to their chemical species both in solution and in particulate matter. For the differentiation of the solid metal species -
e.g., cation exchangeable forms, carbonate phases, reducible fractions,
associations with organic substances and sulfides, and the inert "residual" fractions - chemical extraction sequences have been developed,
which can be used for (i) assessment of sources by characterization of
typical speciation patterns, (ii) estimation of biological availability
of metal pollutants, (iii) differentiation of geochemical environments,
(iv) evaluation of diagenetic effects, and (v) estimation on the potential remobilization of metals under changing environmental condition.
There is a tendency, that elements introduced with solid waste material
are less stably bound than those in natural systems. Even at relative
small proportions of these materials, therefore, mobilization (and subsequent transfer to biota) of potentially toxic elements by acidity,
complexing agents, or redox changes, may be significantly increased.

INTRODUCTION

The general experience that the environmental behavior and toxicity of an element can only be understood in terms of its actual molecular form led to the introduction of the term **"speciation"**, which is
used in a vague manner both for the operational procedure for determining typical metal species in environmental samples and for describing
the distribution and transformation of such species in various media
[1, 2]. The two major categories in applied research are "analyte

species" and "matrix species" [2]: During chemical analysis the species present in a given volume will often be transformed into a single species for which the analytical instrument is sensitive; species exposed to different matrices change their reactivity, solubility, mobility as well as their bioavailability and toxicity.

Among the criteria to assess which element or elemental species may be of major concern, two questions deserve primary attention [3]:

- Is the element **mobile in geochemical processes** because of either its volatility or its solubility in natural water, so that the effect of geochemical perturbations can propagate through the environment?

- What are the critical pathways by which the most **toxic species** of the element can reach the most sensitive organs? Characterizing mobility of a certain element species at the molecular, organelle and cellular level, with respect to its **"bioavailability"** and **"toxicity"**, requires further insight into the interactions of different metals with complexing ligands at physiological concentrations.

Problems of "speciation" become particularly complex in **heterogenous systems**, e.g. in soils, sediments and aerosol particles; thermodynamic models may give suggestions as to the possible species to expect, but due to the important role of kinetically controlled processes in biogeochemistry, the actual speciation is often different from what can be expected [4]. In **polluted ("stressed") systems** entropy increases and there is a concomitant increase in instability in both the physical and biological context [5]: The greater the stress in the environment the more difficulty in **sample handling** and storage prior to analysis. Many of the analytical techniques are handicapped by disruptive preparation techniques which may alter the chemical speciation of inorganic components or lead to loss of analyte before analysis, e.g. freezing, lyophilization, evaporation, oxidation, changes in pH, light catalyzed reactions, reactions with the sample container, time delays before analysis with biologically active samples, sample contamination, statistically invalid sampling, extraction in close to 100% yield; validation of analytical methodology at least needs comparison authentic samples in the same matrix [5].

On the other hand, it is just the "stressed" system, where action is immediately needed and where for an assessment or prognosis of possible adverse effects the species and their transformations of pollutants have to be evaluated. The following questions have been raised with respect to the mobility and bioavailability of potentially toxic metals in contaminated systems [6]:

(1) How reactive are the metals introduced with **solid materials** from anthropogenic activities (hazardous waste, sewage sludge, atmospheric fallout) in comparison to the natural compounds?

(2) Are the **interactions** of critical metals between solution and solid phases comparable for natural and contaminated systems?

(3) What are the factors and **processes of remobilization** to become particularly effective, when either the solid inputs or the solid/solution interactions lead to weaker bonding of certain metal species in contaminated compared to natural systems?

Once the impact of toxic elements or elemental species has been measured or predicted by direct or indirect methods (e.g., bioassays), a management plan can be formulated which usually includes engineering activities [5]. Examples have been given for the treatment of solid waste materials, where valuable elements are sometimes enriched to the extent of economically feasible recycling, for acid mine effluents and for mercury-polluted sediments. A common aspect of most remedial measures in such highly polluted environments is either to extract toxic species or to reduce their mobilities and transfer rates into biological systems [5].

CYCLING AND MOBILITY OF ELEMENTS AND ELEMENTAL SPECIES

Chemical speciation of an element is at first affected by its **source**, e.g. from natural weathering, industrial processing, use of metal components, leaching from garbage and solid waste dumps, animal and human excretions. In the course of its terrestrial and aquatic cycling varying **physico-chemical conditions** may significantly change an element´s species distribution and its behavior in biogeochemical processes [7].

Release of potentially toxic elements into the environment influences ecosystems in global, regional and local **scales**; these impacts can be studied from different **media**, such as soil, water and biota (Table 1). For assessing both the **rates** of input and **historical evolution** of a certain pollutant on a global and regional scale, analysis of dated ice and sediment cores is particularly useful [8, 9].

Table 1. Perturbation of the geochemical cycles of selected metals by society (examples from [4]).

	Scale of Perturbation global reg. local			Diagnostic Environments	Mobilizing Mechanisms	Critical Pathway
Pb	+	+	+	Ice, Sediment	Volatilization	Air, Food
Al	–	+	–	Water, Soil	Solubilization	Water
Cr	–	–	+	Water, Soil	Solubilization	Water
Hg	(–)	+	+	Fish, Sediment	Alkylation	Food (Air)
Cd	(–)	+	+	Soil, Sediment	Solub., Volat.	Food

Of the elements listed in Table 1 **global** perturbations are most dramatically seen for lead. Changes on a **regional** scale are typically found for aluminium mobilization in soils and waters of low buffer capacity affected by acid precipitation; despite insignificant anthropogenic inputs of Al increased solubility will induce toxic effects on both terrestrial and aquatic biota [10]. Chromium usually represents examples of **local** significance; here, elemental **species** exhibit characteristic differences, in that the hexavalent form is more toxic than the trivalent form. Other elements, such as lead and mercury in Table 1, may be mobilized by the biotic or abiotic formation of **organometallic compounds**. Accumulation of methyl-Hg in seafood, probably the most critical pathway of a metal to humans [3]. For describing typical factors enhancing the mobility of metals in both terrestrial and aquatic environments, the element **cadmium** was selected.

Sources and Critical Pathways of Cadmium in the Environment

Subsequent to the catastrophic event of Itai-Itai disease in the Jintsu River area, which was caused by effluents from mine wastes, numerous detailed investigations on soils and waters have been carried out in many countries. Strong cadmium pollution in **aquatic systems** (without indications of acute toxic effects on humans, however) has been recorded in the Hudson River Estuary, New York (nickel-cadmium battery factory), the Hitachi area near Tokyo (braun tube factory), from Palestine Lake, Indiana (plating industry), Sörfjord/Norway and

Derwent Estuary/Tasmania (smelter emissions) and from the Neckar
River/FRG (pigment factory [11]).

Compared to the relatively few critical situations in aquatic
systems, accumulation of cadmium in **agricultural soils** and its increased uptake by plants is of world-wide concern. The aggravating
situation is demonstrated from the example of Switzerland [12]:

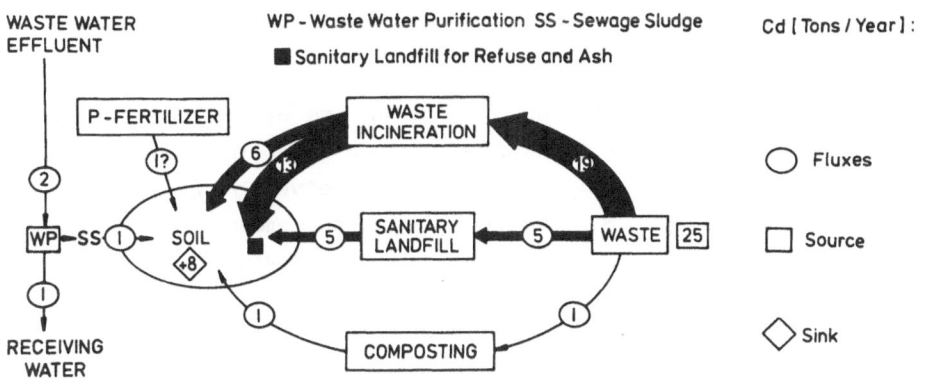

Figure 1.   Cadmium Input into Soil from Different Sources as
            Demonstrated from the Example of Switzerland [12].

It seems that the **atmospheric inputs** of cadmium have been increased approx. 100-fold by anthropogenic activities (including burning of
fossil fuels, smelting of ores, waste incineration, etc.); with these
cadmium levels the upper 30 cm of soil would increase within 170 years
by 0.3 ppm Cd (natural level 0.01-1 ppm). While such inputs are relatively diffuse in character, direct discharges from **municipal waste** materials are playing a particularly critical role for soil contamination
(Figure 1): According to rough estimates 1 t Cd/year is supplied from
both sewage sludge and municipal waste compost, in addition to the
cadmium input via **phosphate fertilizers.** If todays practice of using
compost would be continued, it must be assumed that in 20-30 years a
critical level of 3 ppm Cd in soil would be surpassed, which will prohibit the production of food for human consumption.

Dissolved and Solid Fractions of Cadmium

Mobility of an element in the terrestrial and aquatic environment
is reflected by the ratio of dissolved and solid fractions; these ratios are firstly influenced by the respective inputs and subsequently by

the interactions taking place within the different environmental compartments:

Direct emissions of cadmium into the environment from waste materials are approximately 10-fold higher from solid materials (plated objects, pigments, phosphate fertilizers, sewage sludge, municipal and mining wastes, smelting residues, batteries, coal burning fly ash, etc.) than from dissolved inputs (from active and abandoned lead-zinc mines, sewage treatment, smelter effluents, waste waters from steel works, battery factories and electroplating plants; [13]).

For **atmospheric precipitation** the percentage of dry deposition (by which aerosols or gaseous compounds are deposited on surfaces such as soil particles and plant leaves) has been observed between 10% and 90% of bulk cadmium deposition, depending upon the emission sources, climatic conditions and - in particular - upon pH [14]; the 90% wet deposition is decribed for the situation in West Germany, where the deposited rain water showed pH-values between 3.9 and 4.4 15 . While part of the wet deposition of cadmium is generally adsorbed on organic and inorganic soil constituents, there will be increasing seepage losses when the soil solutions become more acid [16].

For **rivers** it seems that the dissolved fractions of cadmium in polluted waters are significantly higher than in the less polluted systems (Rhine River: 70% [17]; Mississippi River: 10% [18]). This effect could be caused by mobilizing factors such as complexing agents. Relative increases of dissolved metals (Co, Ni, and Cu) in the Susquehanna River during Dec./Jan. and July [19] have been interpreted as an effect of decaying organic matter which is abundant in the river water during these periods [20]. For **lakes** and **oceans** it is suggested that metals from decomposed plankton are partially returned to the solution, whereas the allochthonous particles are deposited on the bottom relatively unchanged [21, 22].

Processes of Solid Metal Speciation: Sorption vs. Precipitation

The situation in rivers should be considered in more detail both for practical and scientific reasons. Where is the more effective remedial action - for dissolved or solid input? Is there an equilibrium between the solid and dissolved phases of a certain element, which can be described by a distribution coefficient "K_D"?

Evaluation of the current literature [23, 24] suggests that there are many factors affecting the **distribution** of trace metals between solution and particulates in aquatic systems:

- **chemical form** of dissolved metal inputs both from natural and civilizational sources [25];

- the **type of interactive processes**, i.e. either sorption/-desorption- or precipitation/dissolution-controlled mechanisms [26];

- concentration and composition of **particulate matter**, mainly with respect to surface-active phases [27] and grain size distribution [28].

Many difficulties are associated with the discrimination of dissolved and particulate metal concentrations by filtration, centrifugation, etc. A characteristic problem is the effect of **colloidal metal species**; colloids seem to have been the "forgotten component in aquatic systems" [29], partly because "few researchers have the facilities or the inclination to investigate these substances".

Two effects relevant for the behaviour of metals both in natural and polluted systems are still not satisfactory explained as yet: One relates to the competition between organic and oxidic adsorption processes, the other is concerned with the discrimination between adsorption/desorption and precipitation/dissolution processes:

With respect to the question of **competition** of various solid substrate constituents, it has been shown by experiments on sorption of radionuclides on reducing sediments [30], that oxidation of samples led to a significant increase in solid-liquid distribution coefficients, whatever the liquid/solid ratio used. This effect is ascribed to the involvement of ferric oxides which are generated in the solid phase and which lead to a displacement of the metal from the humic acid sink. It is suggested, that at high pH, around 9, ferric oxides may be **competitive** with humic acids for metal sorption and that this effect increases with pH [30].

The question of **sorption versus precipitation** has been discussed by Brümmer et al. [31]. It has been suggested, that although in soils, especially in acid once, mainly adsorption-desorption processes of heavy metals combined with complexation processes of organic and inorganic ligands determine the composition of the soil solution, a formation of **definite heavy metal** compounds is also possible under specific conditions. These conditions are relatively high content of the heavy metal concerned, a very low solubility of the metal compound, a sufficient amount of anions and cations needed for the formation of metal compounds and a low content of specific adsorption sites and also of substances (like organic matter) which may prevent the precipitation of definite compounds.

There is a typical **temporal evolution** of the sorption processes, i.e. for those processes, which cannot be explained by a direct precipitation of metals from solution; four different types of evolution (rapid or slower adsorption to nearly 100%; rapid or slow adsorption at a lower level) have been distinguished from experiments using radio-isotopes [32]. These processes are influenced by the hydrological and chemical conditions; sorption of cesium, for example, is typically lowered in the presence of Ca- and Mg-ions. For specific adsorption, **binding strength** typically depends on adsorbend concentration, since there exists a range of site-binding energies [33]: High-energy adsorption sites, since they are fewer in number than lower energy sites, become limiting first; as lower energy sites are gradually filled, the overall binding constant decreases. Particularly in systems containing organic substances, a reduced **reversibility** of metal sorption has been observed [34, 35].

In natural systems, significant differences have been observed for the individual elements, particular with respect to the parameters which are contributing to the limitations of metal concentrations. Data measured by Wollast [36] on River Meuse well correspond to calculated values for lead, copper and zinc, whereas for cadmium the measured concentrations are approximately one order of magnitude lower than predicted from the stability of cadmium carbonate (Table 2):

Table 2: Limitation of dissolved metal concentrations in River Meuse at Tailfer [36].

Element	measured (μg/L)	calculated (μg/L)	from compound:
Lead	6 μg/L	3 μg/L	$Pb_3(CO_3)_2(OH)_2$
Copper	16 μg/L	14 μg/L	$Cu_2(CO_3)(OH)_2$
Zinc	39 μg/L	20 μg/L	$ZnSiO_4$
Cadmium	0.8 μg/L	10 μg/L	$CdCO_3$

21

Figure 2 presents the examples of cadmium and mercury discharges in the River Rhine, which at beginning of 1970's has been known as being extremely polluted in all respects. Meanwhile, considerable change have taken place. In the downstream sections of this river system, at the German/Dutch border, discharges of cadmium have decreased from 250 tons in 1971 to approx. 50 tons in 1983; mercury is even more effectively reduced from 100 tons to approx. 10 tons during this period [37]. This development is due to various factors, such as high water flows in "wet years" and the effects of the economic crisis, particularly at the end of the 1970's [38]. However, a significant portion of the reduction should be affected by improvement of wastewater treatment and by the partial replacement of metals in critical applications.

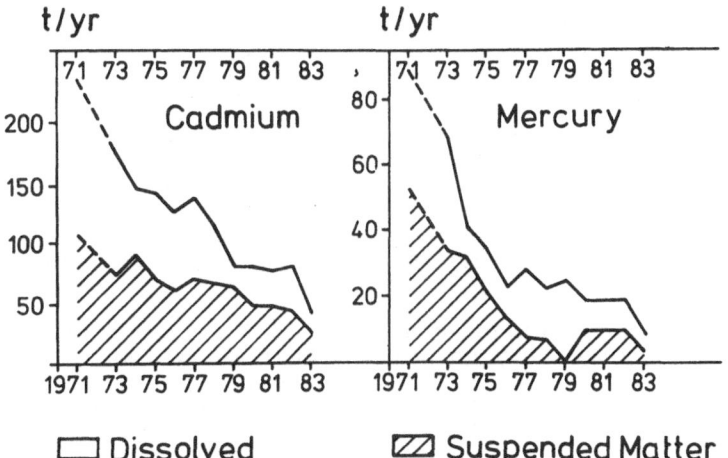

Figure 2: Changes of metal load (dissolved and particulate) in the Rhine River at the Dutch-German border from 1971 to 1983 (after Malle [37]).

It is indicated from these data, that the major decrease for cadmium occurred in the dissolved phase, whereas - until 1979 - the reduction of mercury concentrations mainly took place in the solid phases. At present, therefore, it seems that calculations using **equilibrium data** (i.e. K_D-factors of solid/solution partition coefficients) are inadequate to model natural conditions, because of operational problems, complexity of interactive mechanisms and, in particular, due to the lack of data on **reaction kinetics** of sorption/desorption processes.

ASSESSMENT OF CRITICAL POOLS FOR REMOBILIZATION OF HEAVY METALS

Solid surfaces play an important role in mediating the chemical behaviour of heavy metals; such experience was set in relation to environmental problems by Farmer & Linton [39]: "Accessibility to the environment (via washout, rainout, groundwater leaching, lung fluids, etc.) is governed by both metal **surface accessibility** (extent of surface enrichment) and metal **surface solubility** (surface speciation)". Since adsorption of pollutants onto airborne and waterborne particles is a primary factor in determining the transport, deposition, reactivity, and potential toxicity of these materials, analytical methods should be related to the chemistry of the particle's surface and/or to the metal species highly enriched on the surface. Basically there are **three methodological concepts** for determining the distribution of an element within or among small particles [40, 41]:

- **Analysis of single particles** by X-ray fluorescence using either a scanning electron microscope (SEM) or an electron microprobe can identify differences in the matrix composition between individual particles. The total concentration of the element can be determined as a function of particle size. Other physical fractionation and preconcentration methods include density and magnetic separations.

- The **surface of the particles** can be studied directly by the use of electron microprobe X-ray emission spectrometry (EMP), electron spectroscopy for chemical analysis (ESCA), Auger electron spectroscopy (AES), and secondary ion-mass spectrometry. Depth-profile analysis determines the variation of chemical composition below the original surface.

- **Solvent leaching** - apart from the characterization of the reactivity of specific metals - can provide information on the behaviour of metal pollutants under typical environmental conditions. Common single reagent leachate tests, e.g. U.S. EPA, ASTM, IAEA, ICES, and German Water Chemistry Group (Deutsche Einheitsverfahren) use either distilled water or acetic acid. A large number of test procedures have been designed particularly for soil studies; these partly used organic chelators such as EDTA, DTPA, both as single extractants or in sequential procedures. A single step method of the U.S. EPA [42] designed for studies on the leachability of waste products consists of a mixture of sodium acetate, acetic acid, glycine, pyrogallol, and iron sulfate.

Application of Chemical Extraction Sequences

In connection with the problems arising from the disposal of solid wates, particularly of dredged materials, **extraction sequences** have been applied which are designed to differentiate between the exchangeable, carbonatic, reducible (hydrous Fe/Mn oxides), oxidizable (sulfides and organic phases) and residual fractions [43]. The undisputed advantage of the present approach with respect to the estimation of long-term effects on metal mobilities lies in the fact, that rearrangements of specific solid "phases" can be evaluated prior to the actual remobilisation of certain proportions of the element into the dissolved phase. One of the more widely applied extraction sequence of Tessier and co-workers [44] has been used to examine the different "pools" of Cd and Pb, and to estimate their reactivity in various types of metal-rich particulates (Figure 3).

With respect to the different substrates, the extreme leachability of both cadmium and lead in the **urban particulate matter** [45] and **street dust** [46] is particularly relevant for subsequent interactions with acid, complexing, or salty solutions (the high cation and chloride concentration used in the exchange solution may reflect conditions in soils contaminated with de-icing salt.

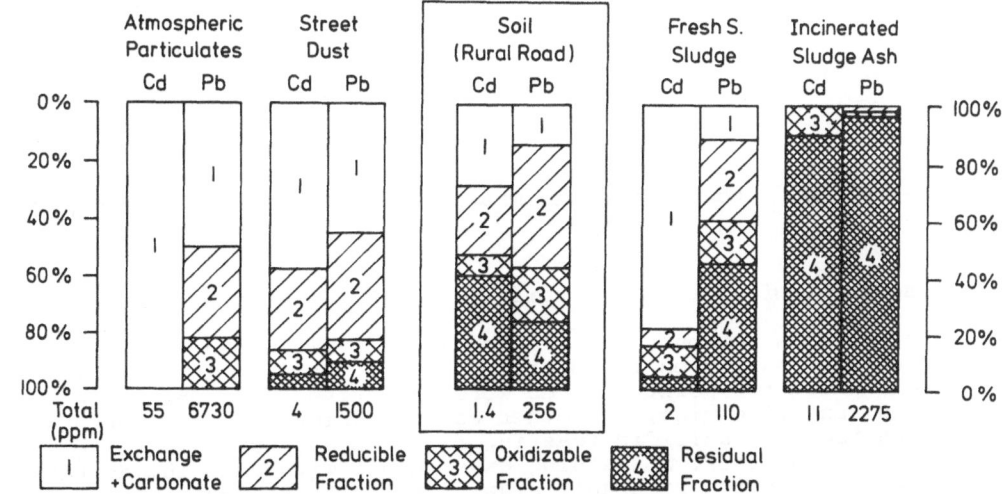

Figure 3. Chemical Fractionation of Cadmium and Lead in Solid
Waste Particles (Urban Particulate Matter Standard
Reference Material (SRM 1648) [45]; street dust [46];
fresh sewage sludge [47]; incinerated sewage sludge
ash [48]). Extractants (after [44], modified for [47]):
(1a) 1 M MgCl$_2$ (Exch.); (1b) 1 M sodium acetate (Carb);
(2) 0.04 M hydroxylamine hydrochloride/25% acetic acid;
(3) 0.02 M nitric acid/30% hydrogen peroxide at pH 2;
(4) conc. nitric acid digestion.

Despite the relatively low concentration of cadium in the **sewage sludge** sample (activated sludge; Landau/Pfalz [47]), there is a significant shift to higher percentages in the carbonate fraction, whereas lead is typically enriched in the organic/sulfidic and residual forms. The general experience that the (anthropogenically) more enriched elements are also the more reactive ones seems to be valid only for waste material, which has not been treated by high temperature processes; this is exemplified by the data from **incinerated sludge ash** (Hamilton municipial incinerator), where the remaining metals are highly enriched but rather stably bound in the "residual fraction" [48]. Better efficiency of emission control, however, of stack gases from combustion processes will recycle higher percentages of atmospheric particulate matter into the residues to be deposited; these materials are expected to contain higher fractions of leachable metals, as shown in the first two columns of Figure 3.

In relation to the species distribution in the **soil** example (Figure 3 represents a moderately polluted soil near a rural road studied by Harrison et al. [46]) the inputs of cadmium and lead from atmospheric particulates, street dust, and fresh sewage sludge are bound in more labile forms and should, therefore, more easily be affected by changes of the chemical environment.

Surface-related studies on solid materials have been performed or are aimed for

(a) assessment of sources by characterization of typical inputs

There are as yet only few studies in this field. We have performed a regional survey on the distribution of thallium in soils in an area, where two point sources were effective [49]: One was an abandoned lead-zinc mine, the other was the chimney of a cement factory which had used sulfidic roasting residues as additives to special cement. Results from leaching experiments with ammonium acetate showed statistically highly significant differences. In the mine area, the extractable portion was in the range of 4%, whereas in the soils affected by the cement plant emissions approx. 18% could be extracted. The lower absolute concentrations in the latter area were therefore more available, leading also in some cases to increased concentrations of thallium in plants.

(b) estimation of biological availability of metal pollutants

Initial applications - about 30 years ago - have been performed in soil science. There is a vast amount of literature [50], and the discussion of this aspect is about the scope of this presentation. However, two recent developments shoud be mentioned. One is the combination of chemical and biological tests (in such an experiment Diks & Allen [51] found a high correlation between the uptake of copper and the amount of copper present in the manganese/easily reducible phase, and it is suggested, that the redox potential and

pH in the gut of the studied worm is such that manganese coatings
are dissolved). Another is the application of body fluids, for
example, for studying the effect of leaching of contaminants from
airborne particles in the lung. It is suggested that biological
chelators, possible cysteine and other serum proteins are important
leaching agents, particularly to remove Zn, V, Cu, and Fe from fly
ash particles in vivo [52].

(c) differentiation of geochemical environments

Diagenesis involves processes in the interstital water and gases,
which are strongly affected by changes of redox conditions. The
sequence of "redox titration" is comprising the major reactions
"respiration", "manganese reduction", "nitrate reduction", "iron
reduction" and "sulfate reduction" mediated by bacteria. In ancient
sedimentary deposits, these zone can be identified from characteristic mineral assemblages [53]: The oxic environment contains oxyhydrates of Mn and Fe at low contents of organic matter (which is
mostly degraded); the post-oxic environment is characterized by the
presence of manganese carbonate, both oxides and carbonates of
iron, and low organic matter as well. Under anoxic conditions two
branches can be distinguished: In the marine milieu reduction of
sulfate provides sufficient sulfide ions to form iron sulfide,
whereas in the freshwater environment there is a tendency to form
carbonate of iron, when the sulfide ions are consumed; the general
tendency of diagenesis in anoxic freshwater sediments is the formation of methane. Sediments collected on four campaigns represent
different early diagenetic environments according to the before-
mentioned classification [54]. Cadmium is associated in the oxic
environment with reducible phases, carbonates and to some extent
with exchangeable forms. Higher percentages of sulfides are found
already in the post-oxic environment and in both anoxic milieus.
These findings reflect the strong affinity of cadmium to the sulfidic phase, even in such environments, where the concentration of
sulfide ions is limited.

(d) evaluation of diagenetic effects

The typical effects of the earliest stages of **"diagenesis"** (involving transformations of organic matter, "aging" of mineral components and formation of new equilibria between solid and dissolved
species) have been demonstrated by Salomons [55] with respect to
the behaviour of trace metals at the sediment/seawater interface.
Desorption was studied by adding cadmium and zinc to suspended matter in river water· after adsorbing periods of 1, 3, 8, 24, and 60
days, NaCl was added to the suspension to increase the chloride
concentration to 1.9% (approx. seawater composition). After an adsorption period of only one day, 24% of the adsorbed cadmium and
60% of the adsorbed zinc remains bound to the sediment; after 60
days 40% of the cadmium and 88% of the zinc bound to the sediment
is not released after NaCl treatment. An extrapolation can be made
to·**geologic time scales** by a comparison of the bonding intensity of
stable metal isotopes and their unstable counterparts - the latter
supplied from radioactive emissions of nuclear power and reprocessing plants. In Figure 4 the effects are shown of sequential leaching of a sediment sample from the lower Rhône River in France. The
reducing agents hydroxylamine (pH 2) and oxalate buffer (pH 3) only
extract 15% of the natural stable manganese while the artificial
isotope Mn-54 from the reprocessing plant is mobilized at more than
80% by these treatments [56].

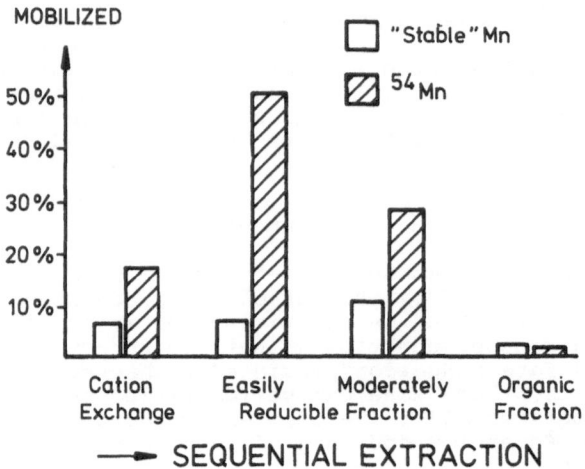

Figure 4: Comparison of Chemical Extractability of Artificial and
Stable Isotopes of Manganese from a Sediment Sample of
the Rhône River [56].

**Estimation of Remobilization Effects under Changing Conditions:
Problems with Sample Pretreatment and Individual Extractants**

Despite of clear advantages of a differentiated analysis over investigations of total sample - sequential chemical extraction is probably
the most useful tool for **predicting long-term adverse effects** from con-
tamined solid material - it has become obvious that there are many problems associated with these procedures [57]:

(a) **Reactions are not selective** and are influenced by the duration of
the experiment and by the ratio of solid matter to volume of extractants. A too high solid content, together with an increased
buffer capacity may cause the system to overload; such an effect is
reflected, for example, by changes of pH-values in time-dependent
tests.

(b) Processes of **readsorption and precipitation** have to be considered,
particularly during extraction with ammonium acetate.

(c) Most important: Labile phases could be **transformed during sample
preparation**, which can occur especially for samples from reducing
environments.

In this respect, earlier warnings have been made by various authors,
not to forget changes of the sample matrix during recovery and treatment of the material. The first relates to the anoxic sediment material, where changes are quite obvious: "The integrity of the samples
must be maintained throughout manipulation and extraction" [43, 58].

The second indicates, that even oxic materials are not safe for changes during treatment: "No storage method completely preserves the initial and physical characteristics even of oxic sediments" [59]. Although these problems, particularly for anoxic sediments, are well known since many years, we have clearly underestimated the effects for a long period. It was realized, when we have been trying to separate iron forms in dredged sediments. Our results were totally different from the model calculations, which claimed for a very considerable percentage of iron carbonate. When we tried to produce artifical iron carbonate, we failed as the originally white material was disintegrating to form red iron oxide within short time periods. It became obvious, that these materials were highly sensitive to aeration.

A simple but impressive experiment on the effect of oxidation in regulating the chemical form of cadmium and other trace metals has been performed on an anoxic sediment sample from Hamburg harbour; the sample was divided into four series under an argon flushed glove box in order to study the effect of various sample pretreatments including aeration and dehydration on the chemical forms of cadmium (Figure 5 [60]):

A Manipulations of the first series were all done under the inert atmosphere to serve as a control.

B The second series was treated by the Elutriate Test modified for air bubbling [61]. This test was initially designed by the U.S. Environmental Protection Agency to detect any short term release of chemical contaminants from polluted material during dredging manipulations and disposal. This test involves the mixing of one volume of the harbour sludge with four volumes of the dredging or disposal site water for a 30-min shaking period. If the soluble chemical constituent in the water exceeds 1.5 times the ambient concentration in the original water, special conditions will govern the remedial measures to be undertaken [62]. A study conducted on the factors influencing the results of the Elutriate Tests has shown that this test as originally developed cannot yield a reliable estimate of the potential release especially of Cd, since it did not define the conditions of mixing to enable a well defined, reproducible oxygen status existing during the test period [61]. The result was that a modified Elutriate Test has been proposed in which compressed air agitation is utilized during the mixing period.

C The third subsample series was freeze dried,

D and the forth series was dried under air in a convection oven at 60°C.

Subsequent to the preservation and pretreatment measures, respectively, the subsamples were extracted by a six-step sequential leaching tech-

nique (modification of Tessier´s method [44]). The significant differences as shown in Figure 5 can be ascribed to the contact of the
sediment with air and by dehydration rather than to experimental arte-
facts such as inhomogeneity of the sediments or variations in the extraction protocol. Indeed, no differences were obtained for oxic suspended matter [63], and the sum of the metal concentrations in the
individual fractions of all four series of each sample agreed within
10%.

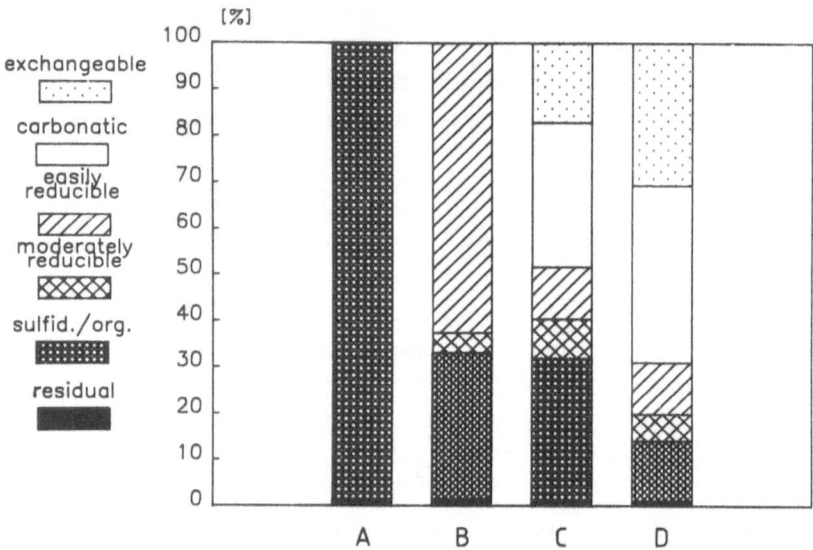

Figure 5: Partition of cadmium in anoxic mud from Hamburg harbour
in relation to the pretreatment procedures: (A) Control
extracted as received under oxygen-free conditions; (B)
after treatment with the Elutriate Test; (C) freeze-
dried; and (D) oven-dried (60°C) [60].

In the pretreatment scheme, which was developed from this experience,
anaerobic dredged samples were taken immediately from the center of the
material with a polyethylene spoon, filled into a polyethylene bottle
until the surface. Immediately after arriving at the laboratory, sediments were inserted into a glove box prepared with an inert argon atmosphere. Oxygen-free conditions in the glove box were maintained by pur-
ging continously with argon under slight positive pressure. Extractants
were deaerated prior to the treatment procedure.

From the application of the various pretreatment procedures, typical behaviour of cadmium in contaminated sediments can be evaluated

[64]: Following the application of the elutriate test, the oxidizable sulfidic/organic portion of Cd decreases drastically and is now found in the easily reducible fraction. Coprecipitation and adsorption of Cd with the precipitated oxyhydrates may have removed the liberated metal from solution. Freshly precipitated oxyhydrates are much more effective in scavenging high concentration of trace metals because of greater reactive surface area than aged crystalline materials [65]. After freeze- and oven-drying of the initially anoxic samples, cadmium proportions were found even in the most mobile operationally defined carbonatic and exchangeable fractions. The high concentration of cadmium present in these fractions may have a hazardous impact on water quality during dredging and disposal operations as well as upland disposal of these sediments [66-68].

REMOBILIZATION OF METALS FROM SOLID WASTE MATERIALS

Solubility, mobility and bioavailability of particle-bound metals can be increased by **four major factors** in terrestrial and aquatic environments:

- **lowering of pH**; acidity imposes problems in all aspects of metal mobilization in the environment: toxicity of drinking water, growth and reproduction of aquatic organisms, increased leaching of nutrients from the soil and the ensuing reduction of soil fertility, increased availability and toxicity of metals, and the undesirable acceleration of mercury methylation in sediments [69]. On a regional scale, acid precipitation is probably the prime factor affecting metal mobility in surface waters;

- increasing occurrence of natural and synthetic **complexing agents**, which can form soluble metal complexes with trace metals that are otherwise adsorbed to solid matter;

- **increasing salt concentrations,** by the effect of competition on sorption sites on solid surfaces and by the formation of soluble chloro-complexes with some trace metals; and

- **changing redox conditions**, e.g. after land deposition of polluted anoxic dredged materials.

Here, particular attention will be given to the effects of **pH and redox changes** in surface waters and soils:

There are as yet only few data on the effect of acid precipitation on metal mobilization in groundwater. It is expected that neutralization of H^+-ions in the unsaturated zone also leads to mobilization

of heavy metals that are discharged into the groundwater together with
sulfate and nitrate ions [70].

In Swedish lakes a pronounced correlation was observed between
dissolved metal levels and pH (Figure 6 after Dickson [71]); this phenomenon is probably due to the **combined effects** of (1) changing solid/-
dissolved equilibria in the atmospheric precipitation, (2) washout processes on soils and rocks in the catchment area, (3) enhancing groundwater mobility of metals and (4) by active remobilization from aquatic
sediments.

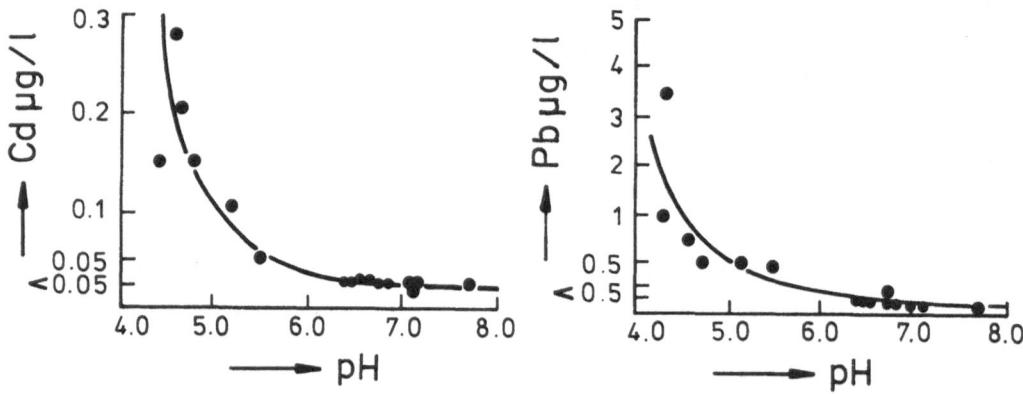

Figure 6. Dissolved Concentrations of Cadmium and Lead in 16
Lakes on the Swedish Coast with Similar Deposition but
with Different pH (after Dickson [71]).

At lower pH-values, one can observe a typical antagonistic behaviour
for cadmium and iron under different **redox-conditions**. An example of
solid waste **compost** in Figure 7a has been studied by Herms & Brümmer
[72]. At pH 5, iron is most mobile at redox values below zero, and concentrations in solution were up to 600 mg Fe/L; cadmium, on the other
hand, has highest dissolved concentrations - up to 1 mg Cd/L - at redox
conditions of +500 mV . It seems, that the study of such **coupled cycles**
will be particularly promising with respect to the interactions of
large redox-controlled systems like iron and manganese with critical
trace metals such as cadmium, mercury and arsenic.

Release of trace metals, particularly of cadmium, from anoxic sludges
has been reported from **estuaries** and coastal marine zones [73]. In
experiments, the major mobilization effects for cadmium occurred after
3-4 weeks treatment time. Therefore, it has been inferred that simple

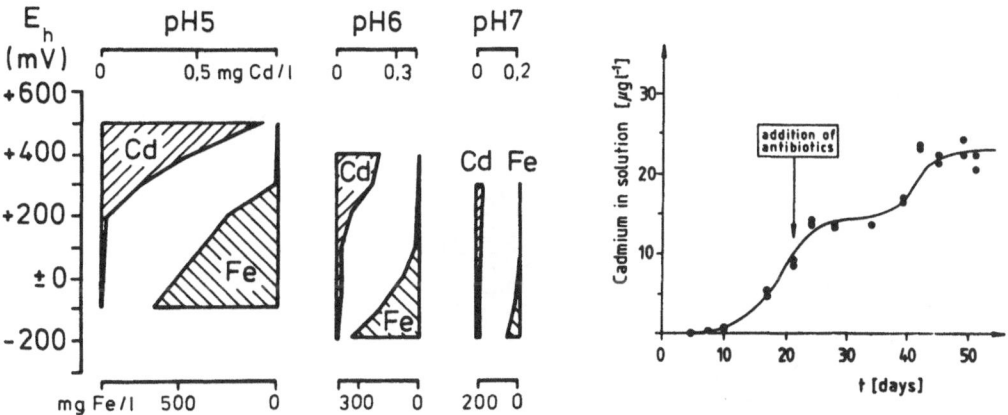

Figure 7. Mechanisms of Cadmium Mobilization from Solid Waste

>   Fig. 7a: Solubility of Cadmium and Iron in Waste Com-
>   post at Different pH and Redox Conditons [72].
>
>   Fig. 7b: Effect of Bacteria on Cadmium Release from
>   Anoxic Sediment under Seawater Conditions [74].

cation exchange mechanisms obviously have no importance [74]. In fact, addition of antibiotics to the system on the 21st day resulted in an interruption of the Cd release for a period similar to the initial lag phase (Figure 7b); it is suggested that bacterial activity can stimulate Cd remobilization by a complex system of substrate decomposition [74].

Mechanisms such as oxidation of organic and sulfidic materials may explain the findings of Hunt & Smith [75] from enclosure experiments in the Narragansett Bay, where the anthropogenic proportion of cadmium in marine sediments is released to the water within approx. three years time; for the remobilization of copper and lead approx. 40 and 440 years is needed, respectively, according to these extrapolations. These data demonstrate the problematic effect of dispersing **anoxic waste materials** in ecologically productive, high-energy nearshore, estuarine, and inlet zones [67], and it has been emphasized by Berman & Bartha [76], that the "contaminated dredged spoils should be prevented from weathering and should be speedily entombed in an anoxic sulfide-rich environment".

An example of oxidative remobilization of cadmium and other heavy metals has been studied in a **tidal freshwater flat** in the upper Elbe estuary near Hamburg [77]. This mudflat - diurnal tidal water fluctua-

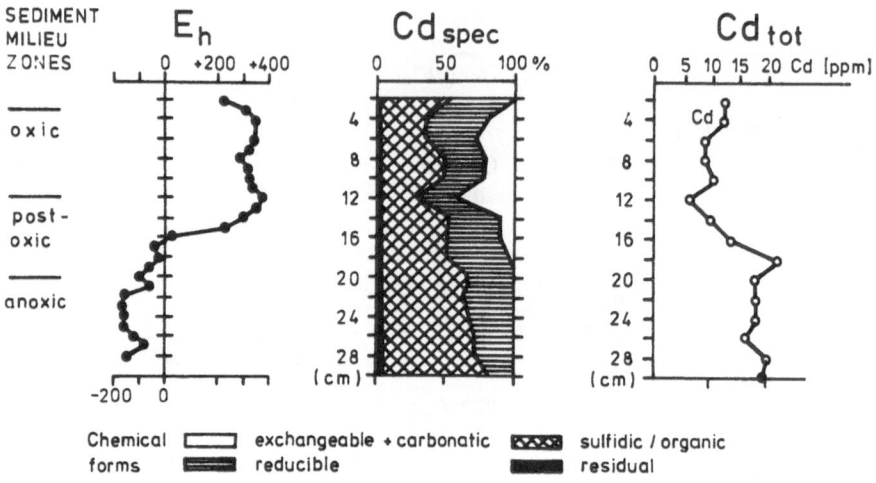

Figure 8. Core Sediments from the Heuckenlock Intertidal Flat
in the Elbe near Hamburg [77]. Left: Sediment Milieu
Zones/Eh-Conditions. Middle: Chemical Forms of Fe and
Cd in Sediment. Right: Bulk Cd Distribution.

tions in the range of 3 m are affecting this productive site - is colo-
nialized by dense monodominant reed stands providing an effective trap
for heavy metal loaden suspended matter from upstreams. Elevated cadmium contents in the rhizomes of the emerged macrophytes indicate high
proportion of bioavailable Cd species in the root zone.

Examination of sediment cores taken at this site showed distinct
pattern of redox potential and heavy metal **fractionation profiles** (Figure 8). Below a marked redox potential discontinuity the Fe fractionation profile is characterized by increasing Na-acetate extractable proportions, which presumably represents carbonatic Fe(II)-forms. The particulate cadmium binding forms, on the other hand, reveal a behaviour
invers to that of iron: While in the anoxic zone approximately 60% to
80% of Cd is found in the oxidizable fraction, high percentages of Na-
acetate extractable forms are found in the oxic and post-oxic zones of
the sediment cores. The higher amounts of labile cadmium forms are
accompanied with a marked depletion in the total content of the toxic
metal compared to that in the anoxic sediment zone. Comparison of the
fractionation patterns and total contents of other diagenetically less
mobile metal examples indicates that a significant proportion of cadmium is leached from the surface sediment by a process of **"oxidative
pumping"** by tidal water drainige in this high-energetic environment.

This could result in migration of the remobilized metal into either the deeper anoxic zone, where it can precipitate again to contribute to the enhanced oxidizable sulfidic/organic fraction, or to the surface water, from where it can be exported into the outer estuary. It could, however, also contribute to bioavailable cadmium portions such as indicated by the enhanced macrophyte cadmium concentrations.

Predicting Potential Metal Mobilization from Solid Speciation

Problems with metals which are dispersed in the environment still exist both on a local and regional scale. Large quantities of **waste materials** on land and in aquatic systems represent long-term reservoirs for the release of metals potentially involving detrimental effects on the quality of both groundwater and agricultural products.

Although the individual "fraction" received from the **chemical leaching experiments** rarely reflect specific metal "phases", but rather are defined by the selection of the extracting medium and by the experimental conditions ("operational phases"), the elution medium is designed to simulate certain - mostly extreme - environmental condition, such as the interaction with saline waters in estuaries or reducing conditions during land disposal of sludge materials. The overall most significant effects on the remobilisation of heavy metals can be expected from acidification, either from atmospheric emissions or from oxidation of sulfidic compounds in anoxic waste materials. Experiments on the long-term behaviour of sludge-induced metals in agricultural soils should pose, therefore, particular attention on changes of redox- and pH-conditions. In addition, the effects of organic substances have to be studied, since the mobility of released metals can be affected by complexation processes.

Species differentiations can be used for the estimation on the remobilisation of metals under **changing environmental conditions:**

- In the **estuarine environment** the "exchangeable fraction" might be affected in particular; however, changes of pH and redox potential could also influence other easily extractable phases, e.g. carbonate and manganese oxides. The high proportion of Cd in exchangeable fractions of polluted sediments is reflected in the typical mobilization effect of this element in the estuarine environment.

- Among the factors enhancing metal mobility **acid interactions** - both from acid precipitation and oxidation of sulfide minerals in mine wastes and dredged sediments - deserve particular attention due to

the fact that ionic species predominates, which is readily available for biological uptake. Lowering of pH will affect, according to
its strength, the "exchangeable", then the "easily reducible" and
in case part of the "moderately reducible" fraction, the latter
consisting of Fe-oxyhydrates in less crystallized forms. In strongly reducing environments, e.g., in highly polluted sediments, the
"moderately reducible fraction" can be affected too, especially
when the iron is present in the form of "coatings".

- The effects on **organically bound metals** are more complex; however
  it has been argued that this fraction is highly susceptible to environmental changes, especially during early diagenetic reactions,
  where recycling of mineralized organic matter and pore fluid transfer processes are controlling the dynamics of pollutants and nutrients in solid waste materials.

Estimations on the long-term behaviour of metals in solid waste materials should pose particular attention on **changes of redox- and pH-conditions.** For determining both effects - short-term release and long-
term rearrangement - we have modified an experimental scheme, which has
originally been used by Patrick et al. [78] and Herms and Brummer [72]
for the study of soil suspensions and municipal waste materials. Chemical fractions are analysed with a six-step extraction sequence before
and after treatment and the released metals are recovered with an ion-
exchanger system from solutions. The variables applied in the **self-
regulating system** are "pH", "redox" and "temperature" (the latter for
simulating the effect of time). The system can be applied for different
intensities of contact between solid materials and solution, e.g. by
using shakers, flow-through columns and dialysis bags.

Predicting Bioavailability of Metals

Bioavailability is a quantitative measure of the **utilization** of an
element under specific conditions [79], and includes mechanisms such as
absorption, transport to a site of metabolic/toxic activity, biotrans-
formation to a metabolically active/toxic form, retention/accumulation,
and excretion [80].

With respect to the **aqueous metal species** it has been suggested
that the "free" or aquo-metal ion form is the most available for
organisms compared to the particulate, complexed, or chelated forms
("equilibrium" or "speciation" hypothesis [81]). Another hypothesis
holds that the kinetics of the **dissociation** of trace metal complexes
[82] - which can be determined using a controlled flow rate column

technique with cation exchange resin Chelex 100 83 - may regulate the toxicity of trace metal, since rapidly dissociating trace metal complexes would constitute the bioavailable fraction of a given metal's total concentration [83]. It has been shown by Engel and Fowler [84] that mobilization of metals by increasing salinity or elevated contents of chelating agents does not necessarily lead to increased accumulation and toxic response of organisms. Recent data of Florence & Stauber [85] for Cu indicate that water-soluble ligands generally decrease the toxicity, whereas lipid-soluble copper complexes were highly toxic.

The present state of knowledge on the interrelation between **solid matter speciation** and the quantitative extent of bioavailable element concentration is still unsatisfactory, since the leachable fraction does not necessarily correspond to the amount available to biota [86]. In most cases information is lacking about the specific mechanism by which the organisms activity participate in the removal of nutrients from the solid substrate. Plant roots activities, for example, include redox changes, pH alterations and organic complexing processes. Competition exists between adsorption sites on the solid substrate (Fe/Mn-oxides, organic matter) and selective mechanisms of metal uptake by the different organisms [87].

Two possible ways for **improving the correlation** of data between the different solid substrates and the organisms (or part of them) could be: (i) Study of soil solution/interstitial water chemistry [88], and (ii) combination of chemical extraction and bioassay [89].

The application of various bioassays to assess the potential pollution arising from dredging activities, as required by legislative mandate in the U.S., has been reviewed by Engler [62]:

(i) Liquid-phase bioassay. Algae and zooplankton were found to respond adequately and may be used to assess stimulation or toxicity. An estimation of mixing and dilution factors that are expected to occur on disposal must be included in the experimental design to simulate field conditions. The mortality of organisms, rather than sublethal considerations, has to be chosen as the indicator of potential environmental effects.

(ii) Solid-phase bioassay. The greatest potential for impact on benthic organisms generally lies with settleable or solid-phase material. When selecting organisms, one filter-feeding, one deposit-feeding and one burrowing species should be included [90]. Mortality is chosen as the interpretative endpoint because of its clear environmental significance.

(iii) Bioaccumulation. Because of long-term effects of bioaccumulation and the short-term nature of the laboratory bioassays (10-day duration), field evaluation of bioaccumulation in site-specific aquatic organisms should be used wherever there has been a historic precedence of disposal at athe site in question. A valid conclusion regarding bioaccumulation is based on statistical significance (95% confidence level) differences in the body burden of specific constituents between organisms in the dump site and the same species living on uncontaminated sediments of similar sedimentologic characteristics.

An integrated approach correlating **bioassay** data of a two-chamber exchange system to the results from a **six-step chemical extraction sequence** was designed by Ahlf [91]. With this system which separates an algal population (or algal cell walls) from suspended solids by a 0.45 fm membrane filter, variations of pH, salinity, redox and effects of complexing agents can be studied for their effects on both accumulation and toxicity.

OUTLOOK: REMEDIAL OPTIONS

Various **remedial measures** against excessive release of toxic metals into the environment may be considered, giving priority to the technologies applied near the source of pollution. If metals have been dispersed, liming of soils and waters as well as mechanical and chemical **stabilization** of solid waste such as encapsulation, use of impermeable base liners and surface covers could reduce fluxes and biological availability of toxic metals [92].

Generally, maintenance of a **pH of neutrality** or slighly beyond (by application of lime or limestone) favors adsorption or precipitation of soluble metals [93, 94]. For **mercury contaminated sediments** isolating from the waterbody by means of physical barriers, such as polymer film overlays, blanket plugs of waste wool, sand and gravel overlays has been proposed [95-97]. Methods of controlling the problem of **acidic mine drainage** include thermodynamic measures (elimination of oxygen and the maintenance of reducing conditions; e.g. by application of sewage sludge on the surface of the spoil heaps) and kinetic effects (e.g., changes of the bacterial propagation cycle [98]). Control of tailing effluents includes seepage collection and handling, and underwater disposal [99]; there are many possibilities of applying physico-chemical methods to **effluent processing**, as presently being applied in other fields of metallurgy (e.g., electroplating).

The best strategy for disposing contaminated sediments is to iso-
late them in a permanently reducing environment [100], e.g. in capped
mound deposits above the prevailing sea-floor or (capped) disposal in
subaqueous depressions; from a geochemical view the marine sulfidic
environment is favourable due to the high stability of metal sulfides,
particularly of mercury (disproportionation of highly toxic monomethyl-
mercury), and more efficient degradation of organic matter [101].

Acklnowledgements

The present review has been improved by discussions with many collea-
gues. In particular, I would like to thank Mr. Michael Kersten for his
suggestions and for providing material on redox effects on metal speciation of sediments. The invitation by the Workshop Committee is
gratefully appreciated.

REFERENCES

1. Bernhard, M., F.E. Brinckman, and P.S. Sadler, Eds. The Im-
    portance of Chemical Speciation in Environmental Processes.
    Dahlem Konferenzen, September 2-7, 1984 Berlin. (Berlin:
    Springer, 1986).

2. Leppard, G.G., Ed. Trace Element Speciation in Surface Waters and
    Its Ecological Implications. Proc. NATO Advanced Research Work-
    shop, Nov. 2-4, 1981 Nervi/Italy. (New York: Plenum Press, 1983).

3. Nriagu, J.O., Ed. Changing Metal Cycles and Human Health. Dahlem
    Konferenzen, March 20-25, 1983 in Berlin. Life Sciences Research
    Report 28. (Berlin: Springer Verlag, 1984).

4. Andreae, M.O., T. Asami, K.K. Bertine, P.E. Buat-Menard, R.A.
    Duce, Z. Filip, U. Förstner, E.D. Goldberg, H. Heinrichs, A.B.
    Jernelov, J.M. Pacyna, I. Thornton, H.J. Tobschall, W.H. Zoller.
    Changing Biogeochemical Cycles, Group Report [3]. pp. 359-374.

5. Wood, J.M., A.M. Chakrabarty, P.J. Craig, U. Förstner, B.A.
    Fowler, U. Herms, I.S. Krull, D. Mackay, G.J. Olson, D.H.
    Russell, W. Salomons, S. Silver. Speciation in Systems under
    Stress, Group Report [1]. pp. 425 - 441

6. Förstner, U. In: H. Bolt et al., Eds. Interactions at the Soil
    Colloid - Soil Solution Interface. Proc. ISSS/NATO Advanced
    Research Workshop, August 25-29, 1986 at Gent/Belgium.

7. Salomons, W., and U. Förstner. Metals in the Hydrocycle. (Berlin:
    Springer Verlag, 1984).

8. Ng, A., and C.C. Patterson. Geochim. Cosmochim. Acta 45: 2109-
    2121 (1981).

9. Müller, G. In: Proc. Intern. Conf. Heavy Metals in the Environ-
    ment, Amsterdam (Edinburgh: CEP Consultants, 1981), pp. 12-17.

10. Campbell, P.G.C., P.M. Stokes, and J.N. Galloway. In: Proceedings International Conference Heavy Metals in the Environment, Heidelberg (Edinburgh: CEP Consultants, 1983), pp. 760-763.

11. Förstner, U. In: Handbook of Environmental Chemistry, Vol. 3/Part A, O. Hutzinger, Ed. (Berlin: Springer Verlag, 1980), p. 59.

12. Keller, L., and P.H. Brunner. Ecotox. Env. Saf. 7: 141-150 (1983).

13. Yost, K.J. In Cadmium-79, Edited Proc. Second Intern. Cadmium Conference, Cannes (London: Metal Bulletin, 1980), pp. 11-20.

14. Nriagu, J.O. In: Cadmium in the Environment, Part 1, Ecological Cycling, J.O. Nriagu, Ed. (New York: John Wiley & Sons, Inc., 1980), p. 71.

15. Nürnberg, H.W., P. Valenta, and V.D. Nguyen. In: Proceedings International Conference Heavy Metals in the Environment, Heidelberg (Edinburgh: CEP Consultants, 1983), pp. 115-123.

16. Mayer, R. Göttinger Bodenkundliche Berichte 70: 1-152 (1981).

17. Heinrichs, H. "Die Untersuchung von Gesteinen und Gewässern auf Cd, Sb, Hg, Tl, Pb, und Bi mit der flammenlosen Atomabsorptions-Spektralphotometrie." Doctoral Dissertation, Universitat Göttingen, Göttingen/FRG (1975).

18. Trefry, J.H. "The Transport of Heavy Metals by the Mississippi River and their Fate in the Gulf of Mexico." PhD Dissertation, Texas A&M University, Dallas (1977).

19. Carpenter, J.H., W.L. Bradford, and V. Grant. In: Estuarine Research, Vol.1, L.E. Cronin, Ed. (New York: Academic Press, 1975), p. 137.

20. Troup, B.N., and O.P. Bricker. In: Marine Chemistry in the Coastal Environment, T.M. Church, Ed. (Washington, D.C.: Amer. Chem. Soc., 1975). ACS Symp. Ser. 18: 133-151.

21. Baccini, P. Z. Hydrol. 38: 121-158 (1976).

22. Bruland, K.W. Earth Planet. Sci. Lett. 47: 176-198 (1980).

23. Förstner, U. and W. Salomons. In: [2] pp. 245-273 (1983).

24. Förstner, U. In: W. Bechteler, Ed., Transport of Suspended Solids in Open Channels, Proceedings of Euromech 192, Munich, June 11-15, 1985. (Rotterdam: A.A. Balkema Publ. 1986).

25. Jenne, E.A., Ed. "Chemical Modeling in Aquatic Systems". ACS Symposium Series 93 (Washington, D.C.: American Chemical Society 1979), 914 pp.

26. Salomons, W. Environ. Technol. Letts. 6: 315-326 (1985).

27. Benjamin, M.M., K.L. Hayes, and J.O. Leckie. J. Water Pollut. Control Fed. 54: 1472-1481 (1982).

28. Duursma, E.K., C. Hoede, C.J. Bosch, and D. Eisma. Neth. J. Sea Res. 3: 423-457 (1967); 4: 395-469 (1970); 6: 265-324 (1973).

29. Allan, R.J. "The Role of Particulate Matter in the Fate of Con-taminants in Aquatic Ecosystems. National Water Research Insti-tute, Scientific Series No. 142. (Burlington: CCIW, 1986).

30. Maes, A., and A. Cremers. In: Speciation of Fission and Acti-vation Products in the Environment, R.A. Bulman and J.R. Cooper, Eds., pp. 93-100 (London: Elsevier Applied Science Publ. 1986).

31. Brümmer, G., J. Gerth, and U. Herms. Z. Pflanzenernährung Bodenkunde 149: 382-398 (1986).

32. Schoer, J., and U. Förstner. In: Proc. First Intern. Conf. on Environmental Contamination, London, July 1984, pp. 738-745 (Edinburgh: CEP Consultants, 1984).

33. Leckie, J.O. In: [1] (1986).

34. Lion, L.W., R.S. Altman, and J.O. Leckie. Environ. Sci. Technol. 16: 660-666 (1982).

35. Förstner, U., W. Ahlf, W. Calmano, and C. Sellhorn. In: Proc. First Intern. Conf. on Environmental Contamination, London, July 1984, pp. 567-572 (Edinburgh: CEP Consultants, 1984).

36. Wollast, R. Water Sci. Technol. 14: 107-125 (1982).

37. Malle, K.-G. Z. Wasser Abwasser Forsch. 18: 207-209 (1985).

38. Förstner, U. In: [3] pp. 71-94 (1984).

39. Farmer, M.E., and R.W. Linton. Environ. Sci. Technol. 18: 319-326 (1984).

40. Keyser, T.R., D.F.S. Natusch, C.A. Evans jr., R.W. Linton Environ. Sci. Technol. 12: 768-773 (1978).

41. Förstner, U. In: R. Leschber, R.D. Davis, and P. L'Hermite, Eds., Chemical Methods for Assessing Bio-Available Metals in Sludges and Soils, pp. 1-30. (London: Elsevier Applied Science, 1985).

42. Ham, R.K., M.A. Anderson, R. Standforth, and R. Stegmann. Background Study on the Development of a Standard Leaching Test. EPA-600/2-79-109. Washington D.C.

43. Engler, R.M., J.M. Brannon, J. Rose, and G. Bigham. In: Chemistry of Marine Sediments, T.F. Yen, Ed. (Ann Arbor/Mich.: Ann Arbor Sci. Publ., 1977) p. 163.

44. Tessier, A., P.G.C. Campbell, and M. Bisson. Anal. Chem. 51: 844-851 (1979).

45. Lum, K.R., J.S. Betteridge, and R.R. Macdonald. Environ. Technol. Letts. 3: 57-62 (1982).

46. Harrison, R.M., D.P.H. Laxen, and S.J. Wilson. Environ. Sci. Technol. 15: 1378-1383 (1981).

47. Förstner, U., W. Calmano, K. Conradt, H. Jaksch, C. Schimkus, and J. Schoer. In: Proc. Intern. Conf. Heavy Metals in the Environment, Amsterdam (Edinburgh: CEP Consultants, 1981), pp. 698-704.

48. Fraser, J.L., and K.R. Lum. Env. Sci. Technol. 17: 52-54 (1982).

49. Schoer, J., and U. Nagel. Naturwissenschaften 67: 261 (1980).

50. Luoma, S.N. Sci. Total Environ. 28: 1-22 (1983).

51. Diks, D.M., and H.E. Allen. Bull. Environ. Contam. Toxicol. 30: 37-43 (1983).

52. Harris, W.R., and D. Silberman. Environ. Sci. Technol. 17: 139-145 (1983).

53. Berner, R.A. J. Sediment. Petrol. 51: 359-365 (1981).

54. Kersten, M., and U. Förstner. Mar. Chem. (in press)

55. Salomons, W. Environ. Technol. Letts. 1: 506-517 (1980).

56. Förstner, U., and J. Schoer. Environ. Technol. Letts. 5: 295-306 (1984).

57. Calmano, W., and U. Förstner. Sci. Total Env. 28: 77-90 (1983).

58. Khalid, R.A., W.H. Patrick, and R.P. Gambrell. Estuar. Coastal Mar. Sci. 6: 21-35 (1978).

59. Thomson, E.A., S.N. Luoma, D.J. Cain, and C. Johansson. Water Air Soil Pollut. 74: 215-233 (1980).

60. Kersten, M., and U. Förstner. Water Sci. Technol. 18: 121-130 (1986).

61. Lee, G.F., J.M. Lopez, and M.D. Piwoni. In: P.A. Krenkel, J. Harrison, and J.C. Burdick, Eds., Proc. Special Conf. Dredging and its Environmental Effects, pp. 253-288. (New York: ASCE, 1976).

62. Engler, R.M. In: R.A. Baker, Ed., Contaminants and Sediments, Vol. 1, pp. 143-169. (Ann Arbor: Ann Arbor Sci. Publ., 1980).

63. Kersten, M., and U. Förstner. In: E.T. Degens, S. Kempe, and R. Herrera, Eds., Transport of Carbon and Minerals in Major World Rivers, Vol. 3, pp. 631-645. (Hamburg: Geolog.-Palaontol. Inst. Univ., 1985).

64. Kersten, M., and U. Förstner. In: J.O. Nriagu, Ed., Cadmium in the Aquatic Environment. (New York: Wiley, 1987).

65. Jenne, E.A. In: W. Chappell and K. Petersen, Eds., Symp. on Molybdenum, pp. 425-553. (New York: Marcel Dekker, 1977).

66. Gambrell, R.P., R.A. Khalid, and W.H. Patrick. Disposal Alternatives for Contaminated Dredged Material as a Management Tool to Minimize Environmental Effects. Tech. Report DS-78-8. (Vicksburg-/Miss.: U.S. Army Engineer WES, 1978).

67. Khalid, R.A. In: J.O. Nriagu, Ed., Cadmium in the Environment, Vol. 1, pp. 257-304. (New York: Wiley, 1980).

68. Salomons, W., and U. Förstner, Eds. Management of Mine Tailings and Dredged Materials. (Berlin: Springer-Verlag, 1987).

69. Fagerström, T., and A. Jernelöv. Water Res. 6: 1193-1202 (1972).

70. Hoeks, J. Water Air Soil Pollut. (1986, in press).

71. Dickson, W. In: D. Drablos, and A. Tollan, Eds., Ecological Impact of Acid Precipitation, pp. 75-83 (Oslo-Aas: SNSF-Project, 1980).

72. Herms, U., and G. Brümmer. Mitt. Deutsche Bodenkundl. Ges. 27: 23-34 (1978).

73. Förstner, U. In: Proc. 4th Intern. Wadden Sea Symp. The Role of Organic Matter in the Wadden Sea. Netherlands Institute for Sea Research, Publ. Ser. No. 10-1984, p. 195 (1984).

74. Prause, B., E. Rehm, and M. Schulz-Baldes. Environ. Technol. Letts. 6: 261-266 (1985).

75. Hunt, C.D., and D.L. Smith. Can. J. Fish. Aquat. Sci. 40: 132-142 (1983).

76. Berman, M., and R. Bartha. Environ. Pollut. 11: 41-53 (1986).

77. Kersten, M., U. Förstner, M. Kerner, and H. Kausch. In: Transactions XIII. Congress Intern. Soc. Soil Science, Hamburg, August 1986, Vol. II, pp. 348-349 (1986).

78. Patrick, W.H., B.G. Williams, and J.T. Moraghan. Soil Sci. Soc. Amer. Proc. 37: 331-332 (1973).

79. Spivey Fox, M.R. et al. Cereal Chem. 58: 6-11 (1981).

80. McKenzie, J.M. In [3] pp. 187-198 (1984).

81. Allen, H.E., R.H. Hall, and T.D. Brisbin. Environ. Sci. Technol. 14: 441-443 (1980).

82. Pankow, J.F., and J.J. Morgan. Environ. Sci. Technol. 15: 1306-1313 (1981).

83. Craft, D. Preprint Extended Abstr. Div. Env. Chem. ACS-Meeting New York, April 1986, p. 236-238 (1986).

84. Engel, D.W., and B.A. Fowler. Environ. Health Perspect. 28: 81-88 (1979).

85. Florence, T.M., and J.L. Stauber. Aquat. Toxicol. 8: 11-26 (1986).

86. Pickering, W.F. Critical Review Anal. Chem. 12: 233-266 (1981).

87. Luoma, S.N., and G.W. Bryan. J. Mar. Biol. Assoc. U.K. 58: 793-802.

88. Carignan, R., F. Rapin, and A. Tessier. Geochim. Cosmochim. Acta 49: 2493-2497 (1985).

89. Munawar, M., R.L. Thomas, H. Shear, P. McKee, and A. Mudroch. "An Overview of Sediment-Associated Contaminants and their Bioassessment". Canadian Technical Report of Fisheries and Aquatic Sciences No. 1253 (1984), 136 p.

90. U.S. Environmental Protection Agency/Corps of Engineers. Ecological Evaluation of Proposed Discharge of Dredged Material into Ocean Water. U.S. Army Engineer Waterways Experiment Station, Vicksburg/Miss. (1977).

91. Ahlf, W. Vom Wasser 65: 183-188 (1985).

92. Rulkens, W.H., J.W. Assink, and W.J.Th. Van Gemert. In: M.A. Smith, Ed., Contaminated Land - Reclamation and Treatment, pp. 37-90. (New York: Plenum Press, 1985).

93. Gambrell, R.P., C.N. Reddy, and R.A. Khalid. J. Water Pollut. Control Fed. 55: 1201-1213 (1983).

94. Calmano, W., U. Förstner, M. Kersten, and D. Krause. In: J.W. Assink, and W.J. Van Den Brink, Eds., Contaminated Soil, pp.737-746. (Dordrecht/The Netherlands: Martinus Nijhoff Publ., 1986).

95. Jernelöv, A., and H. Lann. Env. Sci. Technol. 7: 712-718 (1973).

96. Wolery, T.J., and L.J. Walters. Proc. 17th Conf. Great Lakes Res., IAGLR, pp. 235-249 (1974).

97. Reimers, R.S., P.A. Krenkel, M. Eagle, and G. Tragift. In: P.A. Krenkel, Ed., Heavy Metals in the Aquatic Environment, pp. 117-136. (Oxford: Pergamon Press, 1975).

98. Singer, P.C., and W. Stumm. Science 167: 1121-1123 (1970).

99. Andrews, R.D. In: Proc. Intern. Conf. Heavy Metals in the Environment, Toronto, 1975, Vol. II/2, pp. 645-675. (Toronto: University of Toronto, 1975).

100. Kester, D.R., B.H. Ketchum, I.W. Duedall, and P.K. Park, Eds. Dredged-Material Disposal in the Ocean. (New York: Wiley, 1983).

101. Förstner, U., W. Ahlf, W. Calmano, and M. Kersten. In: G. Kullenberg, Ed., The Role of the Oceans as a Waste Disposal Option, pp. 597-615. (Dordrecht: D. Reidel Publ. Co., 1986).

ANALYTICAL TECHNIQUES IN SPECIATION STUDIES

B. Salbu, Department of Chemistry,
University of Oslo, Blindern, 0315 Oslo 3.

ABSTRACT

Natural waters are dispersed multicomponent systems where most
naturally occurring elements are present in micro or trace concentrations, ranging from 1 mg/l to orders of magnitude below.
In addition trace elements, especially multivalent metals may be
present in different physico-chemical forms varying in size,
charge and density. They may be associated with forms ranging
from simple ions and molecules and via hydrolysis products and
polymers form colloids, pseudo-colloids and may be sorped or
incorporated in organic and inorganic material. The presence of
other components in waters will also influence the distribution
pattern of the element in question as transformation processes
occur.

Due to the low concentrations of trace elements and the variety of
components present in natural waters, combined techniques are most
favourable for speciation purposes. Species are therefore fractionated according to physical and/or chemical properties f.inst.
size and charge prior to their measurements. Then analytical
methods having sufficiently low determination limits for the
elements in question are required.

The present paper will discuss requirements which should be met by
techniques favourable for speciation studies of trace elements in
natural waters.

INTRODUCTION

For macrocomponents the chemical behaviour in waters can be derived from macrochemistry while in dilute solutions, phenomena which are negligible for macroconditioning, become significant. The reaction rates may f.inst. decrease as the probability of effective collisions decreases. Furthermore, reaction directions and mechanisms may change as other processes, such as chemistry at phase boundaries (e.g. sorption) and chemistry of colloidal systems may become predominant. These dilution concentration phenomena are well known from radiochemistry (1-3). Trace elements are therefore defined as elements present in a concentration range where these secondary reaction phenomena no longer can be neglected. These reactions may become significant for a 10^{-5} M solution corresponding to a concentration of 1 mg/l (ppm) for elements having a relative atomic weight of 100 (3). Roughly, 1 ppm can be regarded as the lower limit for which information from macrochemistry is sufficient for describing occurring phenomena.

Physico-chemical states of trace elements.

In natural waters, most trace elements especially multivalent metals, can be present in a variety of physico-chemical forms. As illustrated in Figure 1, trace elements may be associated with forms ranging from simple ions and molecules and via hydrolysis products and polymers form colloids and pseudo-colloids, or be sorped on and incorporated in suspended inorganic or organic particles (4). The dimentions given in the figure are only tentative as the borderline between categories is difficult to establish because gradual transitions occur. The kinetics and the reversibility of the transition processes are, however, still questions to be answered in the future.

The categories indicated in Figure 1 comprise species behaving chemically rather differently as neutral and charged species are not differenciated. Furthermore, the presence of other components in the water will influence the size distribution pattern for an element of interest.

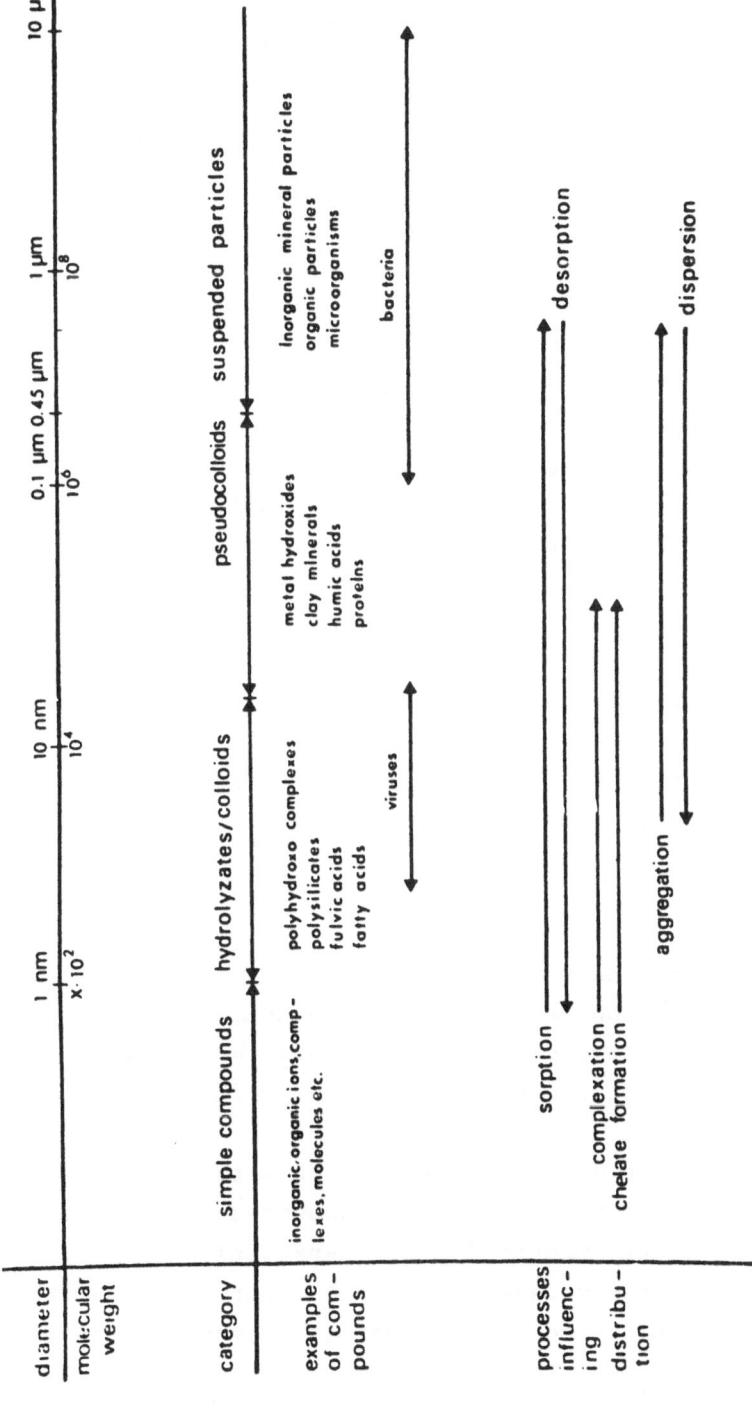

Figure 1 . Association of trace elements with compounds in different size ranges. Transformation processes are indicated.

By sorption to sites offered by foreign surfaces, the low
molecular fraction is usually reduced while f.inst. the pseudo-
colloidal fraction increases. The presence of complexing agents
will often reduce the low molecular fraction while the polymeric
fraction increases. Various factors (e.g. ionic strength) may
influence the stability of colloids and aggregation/dispersion
processes may result in an alteration of the distribution of
species.

TOTAL CONCENTRATIONS

When total concentrations of trace elements in natural waters are
to be determined, the analytical results may depend on the method
used, i.e. its capability to include all species, also the parti-
culate/colloidal fraction, in the analysis.

Based on acidified samples (pH 1, HNO_3, 3 months), scattered
results for total concentrations of Al, Fe, Mn and also Zn were
obtained when using INAA, AAS and ICP (5). Good agreement was
obtained, however, when particles/colloids were removed prior to
analysis. Furthermore, the total concentrations f.inst. for Mn
determined by ICP or AAS were equal to those in fractionated
samples. Thus, a discrimination of particles/colloids and associated elements seems to take place when ICP and AAS are used.

Unless a full decomposition of particles/colloids is obtained, the
results will depend on the preanalysis handling procedure (e.g.
acid conservation, time of storage). Thus, acid conservation may
not be sufficient for decomposing particles retainable by filters.

This particle effect has been observed in waters of different
composition and origin, being more pronounced if inorganic
material is present. Therefore, discrepancies seen in literature
data concerning trace metals in natural waters when different
methods are applied on acid conserved samples only, may at least
partially be attributed to differences in species included in the
analysis.

FRACTIONATION TECHNIQUES

Direct methods are not sufficiently sensitive or selective to obtain information on trace element species in multicomponent heterogeneous systems as natural waters. Therefore, combined techniques where samples are fractionated according to physical and/or chemical properties prior to the analysis itself, are required.

It is quite obvious that different fractionation techniques can be interfaced with molecule, element or isotop specific detectors. Within inorganic chemistry size and charge fractionation techniques are frequently combined with element specific detectors, esp. atomic absorption spectrometry (AAS) and atomic emission inductively coupled plasma spectrometry (ICP). Within organic chemistry, however, combinations such as gas chromatography-mass spectrometry (GC-MS) and high pressure liquid chromatography-mass spectrometry (HPLC-MS) are used for speciation purposes.

For organo-metallic compounds the combination of GC-GFAAS (6), HPLC-AAS (7)/HPLC-IPC (8) has been utilized. Several other combinations of fractionation techniques and detectors are potentially interesting for speciation purposes.

Especially, when different fractionation techniques are combined (e.g. size fractionation and charge fractionation techniques as in electrodialysis) prior to analysis, information on physico-chemical forms of trace elements should be improved. Furthermore, when the volume fractionated is large, it is most useful to use several different analytical methods on each fraction to obtain more data.

For studies of trace element species in natural waters, the critical steps influencing analytical results are sampling, handling of samples and the fractionation techniques applied.

Processes influencing the physico-chemical forms of trace elements takes place during storage of water samples. In order to obtain information on species actually present in a certain water system, the fractionation should take place in situ or at least immediately/shortly after sampling. As all fractionation techniques suffer from disadvantages, methodological effects must be controlled and accounted for.

REQUIREMENTS

Requirements which should be met by fractionation techniques for speciation purposes can be summarized as follows:

a) Fractionation in situ or at the site is essential as storage effects are then avoided.

b) The fractionation should be rapid in order to avoid the establishment of equilibria between species retained and to be separated during the fractionation (i.e. the production of species of interest during fractionation).

c) Equipment surface area to sample volume ratio should be small in order to reduce sorption. However, conditioning with a sample aliquot minimizes this effect.

d) The method should not be sensitive to clogging.

e) Stability of colloids should not be disturbed (e.g. aggregation).

f) Aggregates present should not be disrupted by stirring of solutions.

g) No reagents should be added.

h) The contamination risk should be kept low (e.g. closed systems).

i) Techniques providing large volume fractions are favourable as:

1) The determination limits of elements can be lowered by concentrating the samples from large volumes.

2) Further investigations (e.g. different analytical methods, biotests) can be performed.

The degree of which the different most frequently used fractionation techniques meet these requirements is indicated in Tables 1 and 2.

Table 1. Requirements for size fractionation techniques (x-acceptable)

Requirements:	Sed.:	Cent.:	Filt.:	Ultrafiltration Cell:	Ultrafiltration HF:	Dialysis Cell:	Dialysis HF:	Dialysis In situ:	Gel filtration:
Rapid		X	X		X		(X)		
In situ/at the site	X	X	X		X			X	X
Insignificant sorption		(X)			X			X	
Insignificant clogging	X	X			X		X	(X)	
No aggregation	X	X	X		X		X	X	
No stirring effect	X	X	X		X		(X)	X	X
No reagents added	X	X	X	X	X	X	X	X	
Closed system	X	X	X	X	X	X	X	X	X
Large volumes	(X)	X	X		X		(X)		(X)
Other effects or information needed	large particles removed	Information on density needed	closed systems using compressed inert gas		Rejection characteristics needed			clogging at low flow or in growth season	information on molecular weight/ eluation volume needed

Table 2. Requirements for charge fractionation
techniques (x-acceptable)

| Requirements: | Ion exchange | | Extractions: | Electrochemical methods | |
	Membranes:	Resins:		ASV:	Cont.pot.*):
Rapid	X		X	X	
In situ/at the site	X	X	X		
Insignificant sorption			X		
Insignificant clogging			X	X	X
No aggregation	X		(X)	X	X
No stirring effect	X	X	X	X	(X)
No reagents added				(X)	X
Closed system	X	X	X	X	X
Large volumes	X	X			
Other effects or information needed	filter effect	gel filter effect	collodial particles at phase interface. Distribut. coeff. needed, depends on contact time	change of pH	depends on potential
				depends on deposition time, stirring	
Operationally defined					

*) Controlled potential electrolysis

Among size fractionation techniques, in situ dialysis is favourable being essentially a sampling technique (9-11). However, at low flow rate of water masses during growth season, clogging of membranes occurs. Furthermore, episodic changes in the water quality may not be recognized as the diffusion is a slow process. In these cases large membrane (hollow fiber) ultrafiltration is suitable as fractionation can be performed directly at the site (sampling and subsequent fractionation in close systems). Furthermore, sorption can be minimized by conditioning (4, 11). In addition, large volumes for further investigations are obtainable. The benefit of large membrane (hollow fiber) dialysis is that different low molecular weight species can be distinguished from rate measurements (4).

Conventional ultrafiltration (cell), conventional dialysis (cell) and gel-filtration suffer from several disadvantages. Methodological effects may seriously influence on the distribution pattern and results obtained using these techniques, should be handled with great care.

Analytical results obtained using charge fractionation techniques are operationally defined (Table 2), and are to a certain extent influenced by the presence of polymers, colloids/pseudo-colloids and particles. Therefore, when size fractionation is applied prior to charge fractionation, analytical results are easier to interpret.

Among charge fractionation techniques, extraction and ASV are essentially rapid fractionation techniques. However, reagents added may influence species present, and the fractionation may only be partial. Furthermore, analytical results depend on experimental design esp. the experimental times involved (e.g. mixing time, plating time).

When ionic exchange resins or chelating resins are used, the separation is essentially slow and distortion of original distribution patterns may occur. Using columns, analytical results

depend f.inst. of flow rates. Furthermore, the column may also
act as a gel-filtration resin while filtration through ion-
exchange membranes may be affected by clogging. Thus, retained
species cannot be interpreted as originally charged "free" ions
only.

RADIOACTIVE TRACERS

Chemically well defined radioactive tracers are also most useful
for speciation purposes and for investigating microchemical processes affecting substances in "infinite dilute" solutions. This
is due to the ease of detecting radiation; i.e. the very high
sensitivity using radioactive tracers when compared to other
techniques.

Within analytical chemistry, the use of radioactive tracers represents a powerful tool for investigating the applicability of
analytical procedures. By adding chemically well defined radioactive species to solutions, chemical yields of individual species
at different fractionation steps or whole procedures can be determined.

Tracer experiments can also reveal information on the distribution
of traces between different species in homogeneous systems (i.e.
exchange reactions) or between f.inst. solution and solid phases
in heterogeneous systems (e.g. sorption processes). Furthermore,
dynamic tracer studies can be utilized for investigation of microchemical processes affecting the physico-chemical forms of trace
elements. By combining size and charge fractionation techniques,
the transformation of species can be followed and information on
the reaction paths and kinetics involved can be obtained. Information on naturally occurring components influencing the chemistry
of trace elements can also be achieved if dynamic tracer experiments are performed using model solutions containing specific
interferences as of naturally occurring colloids.

CONCLUSIONS

The presence of different physico-chemical forms of trace elements in natural waters influences the transport, distribution and biological uptake. As most direct methods are not sufficiently sensitive and selective, combined techniques involving fractionation and measurements must be applied. In situ fractionation is most favourable as storage effects are avoided.

Several other requirements should also be met by techniques applicable for speciation purposes. Thus, techniques where methodological effects are minimized, controlled and can be accounted for, should be applied. Chemically well defined radioactive tracers are useful tools for investigating methodological effects influencing analytical results and transformation processes affecting physico-chemical forms of trace elements.

ACKNOWLEDGEMENT

The author will thank Professor A. C. Pappas, University of Oslo, for valuable discussions and the Norwegian Research Council for Science and the Humanities for grants provided.

REFERENCES

1. Starik, I.E. Principles of Radiochemistry, Akademy of Science, USSR. (Translation Series AEC-tr-6314, US Atom Energy Com., 1959).

2. Haïssinsky, M. Nuclear Chemistry and its Application. (Mass., Palo Alto, Cal., London: Addison-Wesley Publ. Comp. Inc., Reading, 1964).

3. Beneš, P. and V. Majer. Trace Chemistry of Aqueous Solutions. (Amsterdam, Oxford, New York: Elsevier 1980).

4. Salbu, B. Preconcentration and Fractionation Techniques in the Determination of Trace Elements in Natural Waters - Their Concentration and Physico-chemical Form. (Oslo: University of Oslo, 1984).

5. Salbu, B., H.E. Bjørnstad, N.S. Lindstrøm, E.M. Brevik,
    J.P. Rambœk, J.O. Englund, K.F. Meyer, H. Hovind, P.E. Paus,
    B. Enger and E. Bjerkelund. Anal. Chim. Acta, 167:161
    (1985).

6. Segar, D.A. Anal. Lett. 7:89   (1974).

7. Manahan, S.E. and D.R. Jones. Anal. Lett. 6:745   (1973).

8. Irgolic, K.J., R.A. Stockton and D. Charkabarse.
    Spectrochim. Acta. 38B:437   (1983).

9. Beneš, P. and E. Steinnes. Wat. Res. 8:947   (1974).

10. Beneš, P., E.T. Gjessing and E. Steinnes. Wat. Res. 10:711
    (1976).

11. Salbu, B., H.E. Bjørnstad, N.S. Lindstrøm, E. Lydersen,
    E.M. Brevik, J.P. Rambœk and P.E. Paus. Talanta 32,
    9:907   (1985).

APPROACHES TO METAL SPECIATION ANALYSIS IN NATURAL WATERS

G.M.P. Morrison

Department of Sanitary Engineering, Chalmers University of Technology,
S-412 96 Göteborg, Sweden

Abstract

Approaches to the separation and identification of metal species in natural waters
are discussed.

Dissolved and colloidal metal species may be fractionated, on the basis of physico-
chemical characteristics, by ion exchange, u.v. irradiation, resin adsorption,
solvent extraction or strong acid digestion. Size fractionation techniques include
filtration, centrifugation, dialysis, ultrafiltration and gel filtration chromatography. Suitable detection techniques, either before or after fractionation, are
anodic stripping voltammetry, ion selective electrodes and atomic absorption. The
bioavailable uptake rate of metal species may be determined by Dialysis with Receiving
Resins.

The separation of particulate associated metal species into fractions is best achieved
by a series of sequential chemical extractions. Metals may be partitioned between the
exchangeable fraction (which is considered to be that which is available primarily
and immediately for biological uptake) the carbonate fraction, the hydrous metal
oxide fraction and the organic/residual fraction.

Additional approaches to speciation analysis include mathematical models and the
product approach, as well as complexation capacity and conditional stability constant
determinations.

Metal Speciation Analysis

Metal speciation analysis involves the fractionation of total metal concentration by
physico-chemical methods (Florence 1986). The fractionation of metal species is
recognised as an essential step in the assessment of the potential biological uptake
and toxicity of metals in a water sample. As a consequence total metal concentrations
may soon be replaced in water quality standards by an assessment of the bioavailable
metal fraction.

Metal Species in Natural Waters

Heavy metals in aqueous systems may occur as organic and inorganic complexes of varying sizes, or be associated with colloidal or particulate material of a hetero-geneous nature (Stumm and Brauner 1975, Steinnes 1983). An important problem, which relates to most natural aquatic systems, is the difficulty of distinguishing between dissolved (0-0.8 nm), colloidal (0.8-400 nm) and particulate (>400 nm) species using conventional physical methods, such as filtration.

Dissolved metal species, particularly those that are free and weakly complexed are potentially available to organisms (Morrison et al. 1984a). In addition, certain lipid soluble metal complexes, such as Cu xanthogenate, rapidly diffuse into biomem-branes and are extremely toxic (Florence 1986).

Separation of Metal Fractions in the Dissolved and Colloidal Phase

Preliminary separation and instrumental techniques have been used to fractionate metals. Instruments which respond to certain metal species include Ion Selective Electrodes (ISE) and Differential Pulse Anodic Stripping Voltammetry (DPASV). Prelimi-nary separation techniques include filtration and ion exchange resins. The different approaches are complementary to each other providing a wide range of information on metal speciation.

Anodic Stripping Voltammetry

DPASV is sufficiently sensitive, with a typical detection limit of about 10^{-9}M, for the direct determination of heavy metals in natural waters (Florence 1982a). This analytical technique can distinguish between the eletrochemically available fraction, which may be toxic, and the bound or electrochemically inert fraction, which is less likely to demonstrate toxic properties.

DPASV has commonly been applied to the primary distinction between "labile" and "bound" metals in filtered water samples (Chau and Lum-Shue-Chan 1974, Duinker and Kramer 1977). The normal procedure for estimating the fraction of labile or electro-chemically available metal involves a standard addition analysis of an untreated sample and is therefore dependent on the kinetics of the reactions controlling the assimilation of the metal spike (Whitfield and Turner 1979). However, Florence (1986) avoids metal spike complexation by calibrating using a blank solution containing standards.

Labile metal, as defined by the experimental conditions, includes ionic as well as some weakly complexed metal. Bound metal is identifiable as the non-labile fraction and is typically associated with a variety of organic and inorganic colloidal materials (Batley and Florence 1976).

Ion Selective Electrodes

Metal ISE respond only to the activity of the toxic free (hydrated) metal ion and have therefore attracted widespread interest (Florence and Batley 1980). The limiting factor for ISE is the non-linear response of the elctrode to metal activity below $\cong$ 10^{-6}M, although measurements in the non-linear region can be made with proper calibration (Midgley 1981).

It is important to satisfy the experimental conditions so that only a Cu ion response is obtained, as the response can also be affected by the physical state of the electrode surface and the chemical environment within which the electrode is working (Frazer et al. 1983). Changes in pH and ionic strength can have a dramatic effect on the electrode response and it is therefore necessary to take these into account when using ISE to monitor environmental samples.

Graphite Furnace Atomic Absorption

Graphite Furnace Atomic Absorption is an automated sensitive technique suitable for measuring total metal concentrations (Astruc et al. 1981.) although it can be used, in conjunction with a preliminary separation process, to provide information on metal speciation. Radojevic et al. (1986) have used Gas Chromatography/Atomic Absorption to determine tetraalkyllead and ionic alkyllead species simultaneously.

Radiotracers

Radiotracers can be added to a sample as a means of tracing the species which a heavy metal can form. Attempts to use ionic radiotracers have been hindered by the slow equilibration of the labelled ions with the non-ionic species of the heavy metal in the sample (Benes and Steinnes 1974). Radiotracers can also be used to investigate metal species uptake by living organisms. The uptake and release of $^{109}Cd^{2+}$ by small fish has been followed using whole body counting (John et al. 1986).

Ion Exchange Techniques

Ion exchange resins can be used to provide some indication of bioavailable metal, for example Chelex-100 is known to take up the free and weakly complexed metals in solution (Morrison et al. 1984a). Thiol resins may more closely resemble the uptake of metal ions by organisms; certainly more Cu is removed by this resin (Florence 1982b). Ion exchange separations have been used in speciation schemes to separate cationic from anionic metal species (Ramamoorthy and Morgan 1983), although it has been stated that such separations probably have little biological relevance (Batley 1983).

Chelex-100 Chelating Resin

Chelex-100 has been successfully applied to seawater metal analysis (Riley and Taylor 1968, Florence and Batley 1976). The resin contains an iminodiacetate chelating group, but it is mainly only the ionic form of the metal which is taken up due to the small pore diameter of the resin (1.5 to 3 nm). In addition weakly complexed metals may dissociate and preferentially associate with the Chelex-100. The ionic form retained by the resin represents the more readily bioavailable metal, whilst the organically coated colloidal particles which are excluded are nevertheless considered to be potentially available for metal solubilisation and transport (Morrison et al. 1984a).

Alternatively, Figura and McDuffie (1979) stated that the slow dissociation of metal complexes in solution, rather than molecular exclusion, is the cause of the incomplete retention by Chelex-100 of some trace metals in natural waters. Therefore column and batch techniques provide for different time scales of metal lability for Chelex-100.

In the column technique metal ions are taken up as the water sample passes down the column at a known flow rate (Florence and Batley 1975, Montgomery and Santiago 1978). In the batch technique longer contact times, up to 16-24 hours, are employed enabling full equilibration of resin and sample metals to be reached (Hart and Davies 1977). The method has proved useful in identifying colloidal associated metals on the basis that these are not kinetically labile during the batch experiment (Figura and McDuffie 1980).

In order to maintain stable pH conditions in the sample and to carry out the experiment over a wide range of pH values the Ca form of Chelex-100 is now preferred (Figura and McDuffie 1977) compared to the Na (Hart and Davies 1977) and H (Florence and Batley 1975) forms.

Thiol Chelating Resins

The presence in organisms of chelating sulphydryl groups has been shown to correlate with heavy metal toxicity (Fisher and Price 1981) and on this basis the metal fraction removed by the thiol group is considered to be a realistic estimate of bioavailable metal (Florence 1982b). The transportation of metals across a cell membrane is believed to be dependent on the lipid solubility of metal species. Similarities between metallothioneins and membrane carrier proteins (Cherian and Goyer 1978,

Kojima and Kägi 1978, Lerch 1980, Roesijaldi 1980) suggest that metals may be complexed and transported in association with sulphydryl groups. This process may be simulated using a thiol resin and it has been reported that thiol material generally has a lower affinity for Pb, Cd and Zn, but a higher affinity for Cu, than Chelex-100 (Florence 1982b).

Techniques for the Measurement of Organically Associated Metals

Chemical and physical methods have been used to decompose dissolved organic materials in natural waters. Some organic compounds are highly resistant and require extensive oxidation before the associated heavy metals are released.

Ultra-Violet Irradiation

The decomposition of organic compounds in natural waters by ultra-violet irradiation has been used to release organically associated heavy metals (Florence and Batley 1977, Laxen and Harrison 1981a). The sample, usually 150 to 200 ml, is introduced into quartz tubes and after the addition of a few drops of 30 % H_2O_2, is irradiated for four to eight hours with a medium pressure ultra-violet lamp of between 500 and 1000 W. The method has had most success in seawater analysis because, in freshwaters, Fe hydroxide is released from an organic colloidal coating and co-precipitates the heavy metals (Laxen and Harrison 1981b, Florence 1982a).

Adsorption of Organics by Resins

It is possible to remove lipid soluble (highly toxic) metal organic complexes from water samples using a high surface area divinylbenzene resin (Florence 1982b). The analysis is carried out at pH 4.0 to prevent free metal ion adsorption to the resin. Following adsorption the metal organic complexes can be eluted with methanol and decomposed by wet acid oxidation or determined by the difference of total metal and soluble metal remaining after resin adsorption (Florence 1982b).

Solvent Extraction

Solvent extraction separates metal species on the basis of their polarity (Batley 1983) and may therefore represent those species that are lipid soluble (Florence 1983). A 9:1 hexane-butanol mixture has a similar dielectic constant to the cell membrane lipid bilayer (Batley 1983).

Oxidation by Concentrated Acids

Organic material can be completely oxidised by the addition of suitable concentrated acids. The addition of concentrated nitric and perchloric acids (9:1) followed by warming and evaporation to dryness is usually sufficient to liberate all heavy metals (Morrison 1985). The remaining metal perchlorates are taken up in 1M HNO_3.

Decomposition of Organics by Ozonolysis

The use of ozone to decompose organics in natural waters has been investigated (Laxen and Harrison 1981a) and shown to give an unexpected decrease in the levels of electro-chemically available Pb, possibly due to the precipitation of Pb (IV).

Filtration

A single filtration step, through a 0.4 μm or 0.45 μm filter is often employed as a preliminary separation for the dissolved and particulate phases. However, this separation is complicated by the presence of colloids (Stumm and Brauner 1975). In addition low pressure filtration is essential as rupture of living cells may occur (Batley and Gardner 1977). Laxen and Harrison (1981a) have introduced a speciation scheme using a series of five nuclepore filters ranging in pore size from 12 μm to 0.015 μm. Nuclepore filters act as barrier, rather than depth, filters allowing a very effective cut-off value and few adsorption losses (Sheldon 1972).

Centrifugation

Centrifugation at 3000 rpm for 30 minutes has been used to separate particles smaller than 190 nm (Benes and Steinnes 1975) whilst centrifugation at 40 000 rpm for up to five hours may remove humic substances (Buffle et al. 1978, Steinnes 1983). Centrifugation can be used as a rapid separation technique, although the dependence on both size and density makes an efficient comparison with other techniques difficult.

Dialysis

Dialysis allows the separation of different groups of metal species on the basis of size (Buffle 1981). Typically ultra-pure water in a dialysis bag is allowed to equilibrate with the sample for 24 hours (Guy and Chakrabarti 1976). Under these conditions and with a pore size of one to five nm (nominal molecular weight cut-off value ≅ 1000) it is found that free dissolved and low molecular weight metal species concentration inside and outside the dialysis bag are equal. An alternative approach is to place the bag in situ for 1 to 14 days until equilibrium is reached. However, over longer experimental time periods the membrane function may be affected as a result of the action of micro-organisms or coatings of organic and inorganic compounds (Salbu et al. 1985).

Hart and Davies (1981) have incorporated a semi-continuous process into their speciation scheme in which a dialysis unit was coupled with a Chelex-100 column. This system was found to reduce the time of equilibration to five hours. When dialysis is performed with hollow fibres the diffusion equilibrium for low molecular weight species is reached within one hour (Salbu et al. 1985).

It has been argued that the dialysis process may realistically represent bioavailable metal uptake, as it is also a membrane transfer process. Cox et al. (1984) found that dialysis with a cation exchange membrane gave similar results to a sulphonate ion exchanger, but gave lower results when compared to uptake by Chelex-100 resin. Dialysis could be combined with a receiving chelating resin to imitate the metal uptake process in a biological cell.

Dialysis with Receiving Resins

Metal bioavailability is determined by the type of metal species present and the transfer and uptake processes at the organism/environmental interface. Metals complexed with humic and fulvic materials are in a relatively non-bioavailable form (Sedlacek et al. 1983, Winner 1985) and therefore emphasis has been placed on the free and weakly complexed metal species as being those forms which are most bioavailable (Florence 1983, Batley 1983, Turner 1984). In addition certain organic lipid soluble metal complexes, such as ethyl xanthogenate metal complexes (a common mineral flotation agent) and Cu 8-hydroxyquinolinate (a herbicide), are likely to rapidly penetrate a cell membrane and damage the cell contents (Florence 1983).

The cell membrane itself is a complex bilayer structure comprised of phospholipids and impregnated proteins. Metal transport processes across and onto the membrane include protein co-transport, direct lipid solubility, surface protein binding and aqueous pore passage. However is has been proposed that metal transport across the membrane may be dominated by diffusion (Luoma 1983) which is supported by the correlation between metal concentration and body size in aquatic insects (Smock 1983, Darlington et al. 1986).

As a major direction of speciation studies is the analysis of metals by a process which imitates metal uptake by organisms, Dialysis with Receiving Resins has been developed as a simple chemical model (Morrison 1985). In Dialysis with Receiving Resins a receiver is encapsulated in a dialysis bag and placed, in situ, in the water of interest. A diffusion gradient is established as the metal concentration is maintained near zero inside the dialysis cell by chelation to Chelex-100 resin. The transfer of metal into the model dialysis cell is expressed in terms of a metal uptake rate per unit membranbe surface area and is more meaningful in terms of bioavailable metal than a direct concentration measurement (Morrison 1986).

Ultrafiltration

Metal species size can be compared with typical cut-off values for ultrafilters. The Amicon PM10 filter has a pore size of 2.8 nm and should permit the separation of free metal ions and small organic and inorganic complexes resulting from trace metals associated with humic substances and colloidal species. Laxen and Harrison (1981a) incorporated this single ultrafiltration step in their filtration based scheme.

Dialysis may be preferred to ultrafiltration on the basis of cost, speed of analysis and efficiency of separation (de Mora and Harrison 1983). Hollow fibre ultrafiltration may also overcome many of the problems of conventional ultrafiltration (Salbu et al. 1985).

Gel Filtration Chromatography

For polluted waters gel filtration chromatography may be an alternative method for separating metal into molecular size fractions (Steinberg 1980). Sample or solute flow is retarded in relation to eluant on a column of porous polymeric beads. The large molecules elute first followed by a continuous size spectrum of molecules (de Mora and Harrison 1984). Preconcentration is often required as the small sample volume compared to eluant gives rise to high dilution factors and large blank values.

Qualitative evidence of large humic complexes (<1500 molecular weight) has been shown for Pb in tap water (de Mora and Harrison 1983, 1984). However, the low metal recovery efficiencies from the gel of 6-50 % precluded the use of gel filtration chromatography as a quantitative metal speciation method for tapwater samples.

The Separation of Heavy Metal Fractions in Suspended Solids and Sediments

Although the toxicity of the dissolved phase is higher because of its direct contact with organisms, the non-lithogenic fractions of suspended solids may subsequently release heavy metals into aqueous systems. The fractions studied must therefore reflect all heavy metal species which might have a direct effect on aquatic biota.

The separation of sediment associated metal species into fractions is best achieved by a series of sequential extractions (Tessier et al. 1979). The extraction steps are not species-selective and repeated treatment often gives a further release of metals, especially in the reducible fractions. They can, however, provide valuable information on the mobility and availability of metals in suspended solids (Morrison et al. 1984a), soils, sediments and sludges (Lum and Edgar 1983).

Exchangeable Fraction

The exchangeable fraction is considered to be that which is available primarily and immediately for biological uptake (Morrison et al. 1984a, Morrison et al. 1984b). It may include metals which are weakly attached to the surfaces of either clays or hydrous Fe and Mn oxides or organic coatings. The addition of a high concentration of chloride or acetate provides a competing ligand for the heavy metals. Hence $MgCl_2$ (Tessier et al. 1979, Eisenreich et al. 1980), $BaCl_2$-triethanolamine (Forstner and Patchineelam 1980) and NH_4OAc (Salomons and Forstner 1980) have been proposed as extracting agents, usually at 1M concentrations and at pH 7.0. The use of ammonium acetate as an extractant has been criticised due to its tendency to attack carbonates (Tessier et al. 1979).

Carbonate Fraction

Carbonates in sediments exist as cements and coatings which co-precipitate with heavy metals. A lowering of pH, such as occurs with acidified rain, would dissolve carbonates and release the associated heavy metals. The most accepted extraction method is a sodium acetate/acetic acid (pH 5.0) treatment although some attack on metals weakly bound to hydrous Fe and Mn oxides may occur (Tessier et al. 1979). Forstner and Patchineelam (1980) claim that their acidic cation exchange method is very specific for carbonate associated metals.

Hydrous Iron and Manganese Oxide Fraction

Hydrous Fe and Mn oxides are thought to exist as coatings on particulate surfaces (Davis and Leckie 1978). These coatings have a high capacity for metal adsorption (Gadde and Laitinen 1974). In order to reduce this fraction (Fe(III) to Fe(II), Mn (IV) to Mn(II)), several extractants have been proposed. Dithionite/citrate (Salomons and Forstner 1980), hydroxylamine hydrochloride/acetic acid (Chao 1972, Tessier et al. 1979) and 0.3M HCl (Eisenreich et al. 1980) all appear to have been used successfully.

The metals in this fraction are more strongly bound than the exchangeable or carbonate fraction and occur mainly as surface associated metals (Davis and Leckie 1978) and co-precipitates of hydrous metal oxides. This fraction is unlikely to have any immediate biological impact, but may accumulate in river or estuarine sediments. Subsequently the metals may be released into the water column when a significant drop in pH occurs (Florence and Batley 1980).

Organic Fraction

Much of the organic fraction will be particle coatings of plant derived organic material. Humic and fulvic acid associated metals can be successfully removed using 1M NaOH. However, the use of sodium hydroxide remobilises metals from phosphates and silicates (Forstner and Patchineelam 1980). Strong oxidising agents, such as 0.4M $Na_4P_2O_7$ (Eisenreich et al. 1980) may also attack the crystalline phase and so Tessier et al. (1979) decided on a compromise and used an extraction mixture of hydrogen peroxide and 0.02M HNO_3. However, metals released from organics by peroxide oxidation are readily adsorbed on clays (Eisenreich et al. 1980).

The relatively strong association of organically bound metals ensures that they are unlikely to be bioavailable. However, this fraction may act as an important transportation mechanism and sink for such metals as Pb and Cu, which have high stability constants with organic compounds (Mantoura et al. 1979).

Mathematical Models

All the mathematical models presented to date are based on thermodynamic considerations only. If the total concentrations and interactions of all the major components of the aqueous system are known and perfect equilibrium conditions prevail, then the concentration of each chemical species of a given element can be calculated. The values of the stability constants and the corresponding dissolved and surface free ligand concentrations must be known. The uncertainty over many stability constants, particularly with organic ligands, makes a comparison with real data difficult. Another problem is the difficulty of considering all the ligands which may be available in an aqueous sample to complex metals.

In seawater organic ligands are generally present at low levels (1 mg/l) and are therefore thought to be relatively unimportant. This allows calculations on the basis of homogeneous chemical reactions and precipitations, particularly as open ocean water is so well mixed. Hence mathematical modelling of seawater has compared favourably with analytical work (Millero 1974, Whitfield 1975).

In freshwaters two other important interactions must be taken into account:

a) The adsorption of metals onto particulates (Jenne 1968, Davis and Leckie 1978) and interactions at the suspended solid/water interphase (Westall 1980, Hohl et al. 1980).
b) The presence of organic material. Most of this material is largely uncharacterised, especially fulvic acids which form important complexes with heavy metals.

Organic compounds may be dissolved or present as surface coatings on particulate material (Davis and Leckie 1978). In addition organic colloids should be taken into consideration.

A computer model, taking these factors into account would need to be backed up by a considerable amount of analytical data. Batley (1983) considers it unlikely that one could ever account for the heterogeneous interaction of metal species with the mixed organic and inorganic colloidal and particulate phases which represent a major component of the total metal concentrations found in most natural systems. Hart and Davies (1981) incorporated data from a speciation study, together with general water quality data, into a computer model. In this way computer calculations were used to assist in the interpretation, rather than prediction, of heavy metal speciation.

The Product Approach

Mixing experiments were introduced by Sholkovitz (1976) to determine the composition of removal products (flocculants) due to the mixing of filtered river and sea waters at varying salinities. The results of these experiments showed that rapid flocculation of Fe, Mn, Al, P, organic carbon and humic acids occurs in the estuarine environment. Most of the precipitate is formed within one half hour of mixing and is removed for analysis. A further 24 hours is required before enough precipitate forms for a second determination.

Further studies (Boyle et al. 1977, Sholkovitz et al. 1978) demonstrated the important role of dissolved organic matter. The high molecular weight component of dissolved humic acids (0.1 μm - 0.45 μm filtered), which constitutes a small fraction of river dissolved organic material, is preferentially and rapidly flocculated during estuarine mixing.

Apparent Heavy Metal Complexing Capacities and Conditional Stability Constants

A large proportion of heavy metals in natural waters are believed to exist in complexed or chelated forms with organic ligands (natural or anthropogenic) which therefore control the geochemical transport and bioavailablility of metals in the aquatic environment (Van den Berg and Kramer 1979). Furthermore colloids may complex and transport trace metals in natural waters.

Analysis of the complexation properties of these organic compounds is complicated by the irregular structure and wide range of components present. Experimental methods to investigate the complexing behaviour of organic fractions in natural waters involve a titrimetric procedure in which the ligands are reacted with a suitable metal ion until the end point, equivalent to the complexing capacity, is reached. A technique to detect remaining free metal is therefore required. Voltammetry is the most tested method (Shuman and Woodward 1973, 1977, Shuman and Cromer 1979) although it may measure weakly complexed metals in addition to free metal ion. Potentiometry (Schnitzer and Kahn 1972), ISE (Buffle et al. 1977), solubilisation (Kunkel and Manahan 1973), bioassay (Davey et al. 1973, Gatcher et al. 1978, Gillespie and Vacarro 1978), dialysis (Sterrit and Lester 1984) and ion exchange (Van den Berg and Kramer 1979) are other available techniques.

Complexation analysis is essentially a complexometric titration of metal ion against ligand and, as ligands are titrated sequentially, those with the highest stability constants are complexed first of all (Crosser and Allen 1977). Changes in the titration slope (added metal versus measured free metal ion) are therefore related to the product of the concentration of the ligand times its conditional stability constant.

The conditional stability constant can be calculated on the basis of formation of the complex (equation 1).

$$aM + bL = M_aL_b \qquad K'_{ML} = \frac{[M_aL_b]}{[M]^a[L]^b} \qquad (1)$$

where a and b depend on stoichiometry
L = Ligand
M = Metal
K'_{ML} = Stability Constant

The complexation capacity and conditional stability constant are calculated according to the method proposed by Ruzic (1982), as shown in equation (2).

$$C_F/C_B = \frac{C_F}{C_L + 1/K'_{ML} \cdot C_L} \qquad (2)$$

where C_F = concentration of free metal
C_B = concentration of bound metal

The straight line plot of C_F/C_B against C_F gives a slope of $1/C_L$ and an intercept of $1/K'_{ML} \cdot C_L$.

Sterritt and Lester (1984) have compared the conditional stability constants obtained for Cd, Pb and Cu with fulvic acid by using three different analytical techniques; ISE, dialysis and DPASV. The stability constants obtained by DPASV measurements were lower than for the other two methods, suggesting the existence of increased metal complex lability. However, the complexation capacity was similar for all methods. ISE detected both weak and strong complexes, while DPASV and dialysis were only sensitive to stronger complexes. For stronger complexes a log K range of 4.7 to 7.0 was found, while weaker complexes gave a log K value of between 3.8 and 5.6 for all three metals. Weaker binding sites often show a much greater metal capacity for Cu and Cd than stronger binding sites (Sterritt and Lester 1984, Sterritt and Lester 1985).

Acknowledgements

Financial support from the National Swedish Environmental Research Council is gratefully acknowledged.

This article is to a great extent based on the "Handbook for Metal Speciation in Natural Waters", published internally by the Department of Sanitary Engineering, Chalmers University of Technology, Göteborg, Sweden. Both Mike Revitt, Middlesex Polytechnic and Brit Salbu, University of Oslo made extensive contributions to the handbook.

References

Astruc, M., Lecomte, J., and Mericam, P. (1981). Environmental Technology Letters, 2, 1-8.

Batley, G.E. (1983). In: Leppard, G.G. (ed.), Trace Element Speciation in Surface Water and its Ecological Implications, NATO Conf. Ser., Vol. 6, 17-36, Plenum Press.

Batley, G.E., and Florence, T.M. (1976). Analytical Letters, 9, 279-388.

Batley, G.E., and Gardner, D. (1977). Water Research, 11, 745-756.

Benes, P., and Steinnes, E. (1975). Water Research, 9, 741-749.

Boyle, E.A., Edmond, J.A., and Sholkovitz, E.R. (1977). Geochimica Cosmochimica Acta, 41, 1313-1324.

Buffle, J. (1981). Trends in Analytical Chemistry, 1, 90-95.

Buffle, J., Greter, F., and Haerdi, W. (1977). Analytical Chemistry, 49, 216-222.

Buffle, J., Deladoey, P., and Haerdi, W. (1978). Analytica Chimica Acta, 101, 339-357.

Chao, T.T. (1972), Proceedings of the American Soil Science Society, 36, 764-768.

Chau. Y.K. and Lum-Shue-Chan, K. (1974). Water Research, 8, 383-388.

Cherian, M.G., and Goyer, R.A. (1978). Life Science, 23, 1-10.

Cox, J.A., Slonawski, K., Gatchell, D.K., and Hiebert, A.G. (1984). Analytical Chemistry, 54, 650-653.

Crosser, M.L., and Allen, H.E. (1977). Soil Science, 123, 176-181.

Darlington, S.T., Gower, A.M., and Ebdon, L. (1986). Environmental Technology Letters, 7, 141-146.

Davey, E.W., Morgan, M.J., and Erickson, S.J. (1973). Limnology and Oceanography, 18, 993-997.

Davis, J.A., and Leckie, J.O. (1978). Environmental Science and Technology, 12, 1309-1315.

De Mora, S.J., and Harrison, R.M. (1983). Water Research, 17, 723-733.

De Mora, S.J., and Harrison, R.M. (1984). In: Environmental Contamination, 124-130, CEP Consultants Ltd, Edinburgh.

Duinker, J.C., and Kramer, C.J.M. (1977). Marine Chemistry, 5, 207-228.

Eisenreich, S.J., Hoffman, M.R., Rastetter, D., Yost, E., and Maier, W.J. (1980). In: Kavanaugh, M.C., and Leckie, J.O. (ed.), Particulates in Water, 136-176, ACS Symposium Series No. 189, Amer. Chem, Soc,, Washington D.C.

Figura, P., and McDuffie, B. (1977). Analytical Chemistry, 49, 1950-1953.

Figura, P., and McDuffie, B. (1979). Analytical Chemistry, 51, 120-125.

Figura, P., and McDuffie, B. (1980). Analytical Chemistry, 52, 1433-1439.

Fisher, N.S., and Price, G.J. (1981). Journal of Phycology, 17, 108-111.

Florence, T.M. (1982a). Talanta, 29, 345-364.

Florence, T.M. (1982b). Analytica Chimica Acta, 141, 73-94.

Florence, T.M. (1983). Trends in Analytical Chemistry, 2, 162-166.

Florence, T.M. (1986). Analyst, 111, 489-505.

Florence, T.M., and Batley, G.E. (1976). Talanta, 22, 201-204.

Florence, T.M., and Batley, G.E. (1976). Talanta, 23, 179-186.

Florence, T.M., and Batley, G.E. (1977). Talanta, 24, 151-158.

Florence, T.M., and Batley, G.E. (1980). Critical Reviews in Analytical Chemistry, 219-296.

Forstner, U., and Patchineelam, S.R. (1980). In: Kavanaugh, M.C., and Leckie, J.O. (ed.), 177-183, Particulates in Water, ACS Symposium Series No. 189, Amer.Chem. Soc., Washington D.C.

Frazer, J.W., Balaban, D.J., Brand, H.R., Robinson, G.A., and Lanning, S.M. (1983). Analytical Chemistry, 55, 855-861.

Gadde, F.F., and Laitinen, H.A. (1974). Analytical Chemistry, 46, 2022-2026.

Gatcher, R., Davies, J.S., and Mares, A. (1978). Environmental Science and Technology, 12, 1416-1421.

Gillespie, P.A., and Vaccaro, R.F. (1978). Limnology and Oceanography, 23, 253-548.

Guy, R.D. and Chakrabarti, C.L. (1976). Chemistry in Canada, 54, 26-29.

Hart, B.T., and Davies, S.H.R. (1977). Australian Journal of Marine and Freshwater Research, 28, 397-402.

Hart, B.T., and Davies, S.H.R. (1981). Estuarine and Coastal Shelf Science, 12, 353-374.

Hohl, H., Sigg, L., and Stumm, W. (1980). In: Kavanaugh, M.C., and Leckie, J.O. (ed.), Particulates in Water, 1-31, ACS Symposium Series No. 189, Amer. Chem. Soc., Washington D.C.

Jenne, E.A. (1968). In: Baker R.A. (ed.), Trace Inorganics in Water, 337-387, ACS Symposium Series No. 73. Amer. Chem. Soc., Washington D.C.

John, J., Gjessing, E.T., Grand, M., and Salbu, B. (1986). Proceedings Int. Humic Subst. Soc. Meeting, University of Oslo.

Kojima, Y., and Kägi, J.H.R. (1978). TIBS, 90-93.

Kunkel, R., and Manahan, S.E. (1973). Analytical Chemistry, 45, 1465-1468.

Laxen, D.P.H., and Harrison, R.M. (1981a). Science of the Total Environment, 19, 59-82.

Laxen, D.P.H., and Harrison, R.M. (1981b). Water Research, 15, 1053-1065.

Lerch, K. (1980), Nature, 184, 368-370.

Lum, K.R., and Edgar, D.G. (1983). Analyst, 108, 918-924.

Luoma, S.N. (1983). Science of the Total Environment, 28, 1-22.

Mantoura, R.F.C., Dickson, A., and Riley, J.P. (1978). Estuarine and Coastal Marine Science, 6, 387-408.

Midgley, D. (1981). Ion Selective Electrode Reviews, 3, 43-104.

Millero, F.J. (1974). In: Goldberg, E.D. (ed.), The Sea, Vol. 5, Marine Chemistry, 3-81, Wiley-Interscience, New York.

Montgomery, J.R., and Santiago, R.J. (1978). Estuarine and Coastal Marine Science, 6, 111-116.

Morrison, G.M.P. (1985). Ph.D. Thesis, Middlesex Polytechnic, U.K., 316 pp.

Morrison, G.M.P. (1986). Environmental Technology Letters, in Press.

Morrison, G.M.P., Revitt, D.M., Ellis, J.B., Svensson, G., and Balmēr, P. (1984a). In: Balmēr et al. (ed.) Urban Storm Drainage, Department of Sanitary Engineering, Chalmers Tekniska Högskola, Sweden, Vol. 3, 989-1000.

Morrison, G.M.P., Revitt, D.M., Ellis, J.B., Svensson, G., and Balmēr, P. (1984b). In: Vernier, J.P. (ed.), Interactions between Sediments and Water, CEP Consultants Ltd., Edinburgh, 226-229.

Radojevic, M., Allen, A., Rapsomanikis, S., and Harrison, R.M. (1986). Analytical Chemistry, 58, 658-661.

Ramamoorthy, S., and Morgan, K. (1983). Regulatory Toxicology and Pharmacology, 3, 172-177.

Riley, J.P., and Taylor, D. (1968). Analytica Chimica Acta, 40, 479-485.

Roesijaldi, G. (1980). Marine Environmental Research, 4, 167-179.

Ruzic, I. (1982). Analytica Chimica Acta, 140, 99-113.

Saar, R.A., and Weber, J.H. (1982). Environmental Science and Technology, 16, 510A-517A.

Salbu, B., and Bjørnstad, H.E., Lindstrøm, N.S., Lydersen, E., Brevik, E.M., Ramboek, J.P., and Paus, P.E. (1985). Talanta, 32, 907.

Salomons, W., and Forstner, U. (1980). Environmental Technology Letters, 1, 506-517.

Schnitzer, M., and Kahn, S.U. (1972). Humic Substances in the Environment, Dekker, New York.

Sedlacek, J., Källqvist, T., and Gjessing, E. (1983). In: Christman, R.F., and Gjessing, E. (ed.), Aquatic and Terrestrial Humic Materials, 26, 495-516, Ann Arbor Science, Michigan.

Sheldon, R.W. (1972). Limnology and Oceanography, 17, 494-498.

Sholkovitz, E.R. (1976). Geochimica Cosmochimica Acta, 40, 831-845.

Sholkovitz, E.R., Boyle, E.A., and Price, N.B. (1978). Earth and Planetary Science Letters, 40, 130-136.

Shuman, M.S., and Cromer, J.L. (1979). Environmental Science and Technology, 13, 543-545.

Shuman, M.S., and Woodward, G.P. (1973). Analytical Chemistry, 45, 2032-2035.

Shuman, M.S., and Woodward, G.P. (1977). Environmental Science and Technology, 11, 809-813.

Smock, L.A. (1983). Freshwater Biology, 13, 313-321.

Steinberg, C. (1980). Water Research, 14, 1239-1250.

Steinnes, E. (1983). In: Leppard, G.G. (ed.), Trace Element Speciation in Surface Waters, 37-47, NATO Conf. Ser., Vol, 6., Plenum Press.

Sterritt, R.M., and Lester, J.N. (1984). Water Research, 18, 1149-1153.

Sterritt, R.M., and Lester, J.N. (1985). Water Research, 19, 315-321.

Stumm, W., and Brauner, P.A. (1975). In: Riley, J.P., and Skirrow, G. (ed.), Chemical Oceanography, Vol. 1, 173-239, 2nd edition, Academic Press.

Tessier, A., Campbell, P.G.C., and Bisson, M. (1979). Analytical Chemistry, 51, 844-851.

Turner, D.R. (1984). In: Siegel, H. (ed.), Metal Ions in Biological Systems, Vol. 18, 137-164, Marcel Dekker, New York.

Van den Berg, C.M.G., and Kramer, J.R. (1979). Analytica Chimica Acta, 106, 113-120.

Westall, J. (1980). In: Kavanaugh, M.C., and Leckie, J.O. (ed.). Particulates in Water, 33-44, ACS Symposium Series No. 189, Amer. Chem. Soc., Washington D.C.

Whitfield, M. (1975). In: Riley, J.P., and Skirrow, G. (ed.), Chemical Oceanography, Vol. 1, 44-162, 2nd edition, Academic Press, London,

Whitfield, M., and Turner, D.R. (1979). In: Jenne, E.A. (ed.), Chemical Modelling in Aqueous Systems, 657-680, ACS Symposium Series No. 93, Amer. Chem. Soc., Washington D.C.

Winner, R.W. (1985). Water Research, 19, 449-455.

METAL FRACTIONATION BY DIALYSIS - PROBLEMS AND POSSIBILITIES.

Hans Borg
The National Environmental Protection Board
Trace Metal Laboratory
Box 1302
S-171 25 SOLNA, Sweden.

Abstract

Dialysis is a convenient method to separate different molecular-weight species in natural waters or other solutions. Some applications of dialysis methods for the fractionation of metal forms in water have been reported. The main advantage of dialysis methods are:

- possible to separate truly disolved forms from colloids
- relatively few contamination problems
- suitable for in situ measurement.

As with other fractionation methods there are some problems to be aware of, for example:

- relatively slow diffusion rate
- no definite separation between inorganic and low molecular-weight organic forms
- changed pore size caused by microbial attack and clogging with organic matter.

Results from different applications of in situ dialysis in acidified, limed and metal polluted waters are dicussed.

Introduction

The concentration of trace metals in unpolluted waters are mostly in the sub-microgram per litre level. Only elements like Fe, Mn, Al and often Zn are frequently present in concentrations higher than one microgram per litre, while Cd for instance, generally is found in the range 0.005-0.05 µg/l (1,2). Because of these low concentrations, it is difficult even to perform sampling and analysis of total concentrations accurately. Most fractionation techniques, such as filtration, ultrafiltration or centrifugation increase the contamination problems during the procedure. A technique which is relatively free of these problems is dialysis, especially when performed _in situ_. Some applications of _in situ_ dialysis for the collection of dissolved metals in water have been reported in the literature (3,4,5,6,7,). The method is also possible to combine with chelating resin (8). _In situ_ dialysis has also been used to culture fytoplankton for the evaluation of the effect of pollutants (9).

Methods

The dialysis experiments were performed in lakes in the province of Småland, in the Åva nature reserve south of Stockholm, in the Delsbo area in Hälsingland, in mountain brooks in Lofsdalen, Härjedalen and in lakes around Skellefteå, Västerbotten in northern Sweden. The measurements were generally repeated two times a year (April and September) during 1981 -1986.

The experiments were performed as described in references 6 and 10. The dialysis tubings were filled with ultra-pure (Milli-Q) water and allowed to equilibrate in the lake waters for 5 days. Samples for determination ot total metal concentrations and other water chemical parameters were collected at the start and at the end of the dialysisperiod. The tubings were then brought to the laboratory in bottles filled with the same water as they had been equilibrated in. The outside was cleaned, the top of the tubing was cut off and the content was pipetted into a polyethylen bottle. Ultra-pure conc. nitric acid was added for preservation (2 ml/l). All the steps in the laboratory were performed in a class-100 clean bench. All storage bottles and other laboratory ware were acid cleaned as described in ref. 11.

The metal determinations were performed with flame AAS (Fe,Mn) or graphite furnace AAS (Zn,Cu,Pb,Cd,Al). To increase the accuracy, the determinations of Cd,Pb and Cu were made after a preconcentration of about ten times by freeze drying in the sample bottles.

Results

The results reported here are partly presented in previous reports which will be referred to in the text. The dialysis experiments performed in various lake waters have been able to show some general features of the elements:

Iron occured to a great deal as particulate bound even in relatively acid waters. Generally, just a few percent was dialysable except in very clear waters (Fig.1, ref.12).

Manganese occured more in dialysable form, often to about 90 % . However, the variation was very large depending on pH and especially on the content of organic substances (TOC) (Fig1.).

Lead was generally found as particulate especially at high humic content (Fig.2). In more clear waters there was a larger fraction that was dialysable and negatively correlated to pH (6,10).

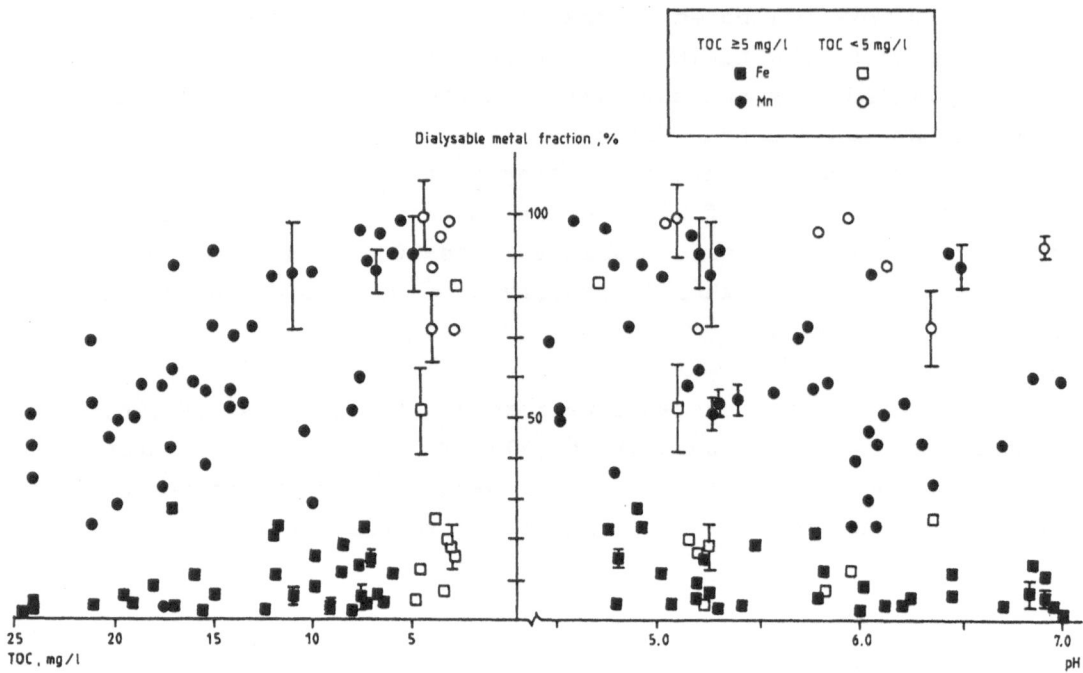

Fig. 1. Dialysable fraction of Fe and Mn in Swedish forest lakes
plotted against pH and TOC. Error bars represent standard
deviations of three to four measurements.

Cadmium and **zinc** existed mainly as dialysable, especially at low pH
values. Even at pH 7 up to around 60% of cadmium was dialysable. No
correlation between dialysable fraction and the TOC concentration
was found for zinc and cadmium (6,10).

Dialysable **copper** showed a negative correlation with TOC but not
with pH (6). Copper was mostly found in the colloidal fraction,
which probably consists of humic substances (10).

Dialysable **aluminium** showed a negative correlation both to pH and
TOC values (6). At pH 5, about 70% of aluminium was dialysable in
clear water lakes (TOC < 5 mg/l).

Arsenic occured to a relatively high degree in dialysable form (30-
60 %). In lakes with low TOC concentration around 90 % of arsenic
was dialysable (10).

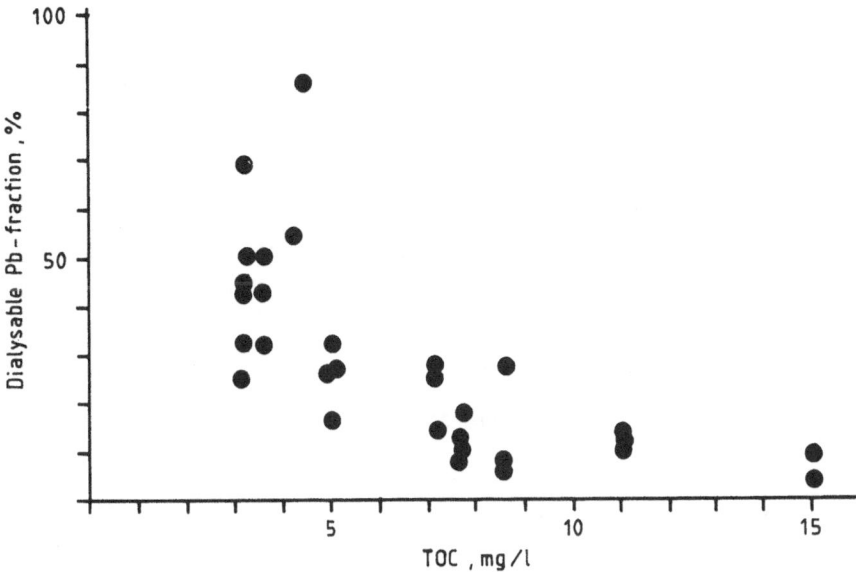

Fig. 2. Influence of TOC on the dialysable Pb fraction in Swedish
forest lakes.

Discussion

For separation of dissolved metals in water, the _in situ_ dialysis
technique have some advantages over filtration techniques. The main
advantages can be summerized as follows:

- separates truly dissolved species
- relatively small contamination problems
- practicly unlimited supply of material to dialyze
- no adsorption losses at equilibrium
- possible to combine with chelating resins
- low cost compared to ultrafiltration.

However, as with all other fractionation techniques, there are also
some problems to be aware of.

The diffusion rate of elements through the membrane is rather slow.
In deionized water, equilibrium is generally reach within 24 hours,
but in natural waters the diffusion is slower,as shown in Fig.3. The
membrane used had a pore-size of 2.4 nm. Similar results were
reported by Beneš and Steinnes (3) who found that the diffusion rate
for organic complexes like Fe-EDTA was slower than for inorganic
ions. The relatively slow diffusion rate might cause problems when
dialysis is used in running waters with fast changes in water
chemistry (12).

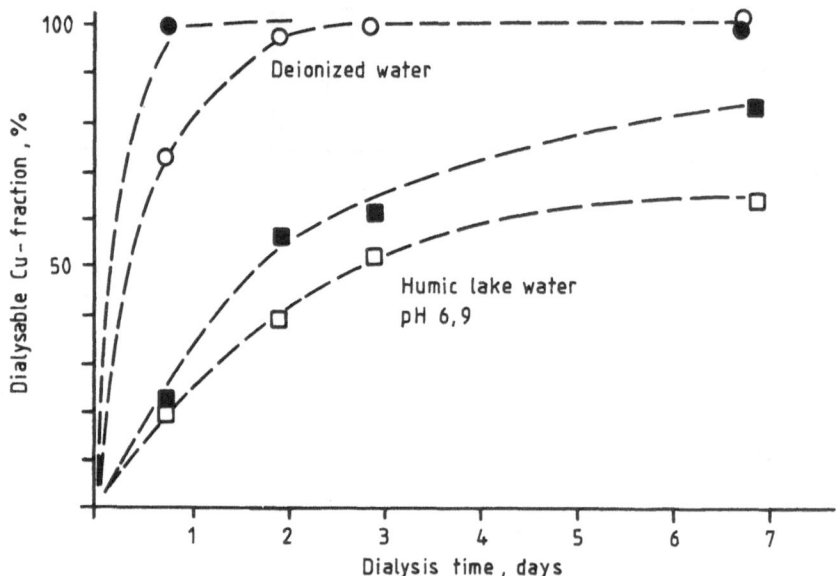

Fig. 3. Percentage of dialysable Cu as a function of time. The
    filled symbols represent samples with additions of sulphate
    (10 meq/l), chloride and bicarbonate (1 meq/l). 2000 mwco
    membrane.

Donnan disequilibrium can occur if the organic anions make up a
substantial fraction of the total equivalents of ions (Bull 1951,
cited in ref.13). Then it will be more inorganic anions inside and
more cations outside then would normally occur. However, according
to calculations by LaZerte (13) who applied a correction factor for
the Donnan effect when measuring dissolved Al, relatively high TOC-
values (around 10 mg/l or more) is required to cause any significant
error.

Depending on the molecular-weight cut-off of the membrane, some
organic compounds will also penetrate. Fig.4 shows the organic
penetration in lake waters with different TOC contents, measured
inside and outside the dialysis tubing. The membrane had a mwco of
about 2000 and the dialysis time was 7 days. LaZerte (13) reported
an organic penetration of 6-12 % even in a 1000 mwco membrane for a
dialysis time of 24 h. If these findings mean that organically
complexed metal forms will be included in the dialysate is still
unclear, but that might be expected as the low molecular-weight
organic fractions passing through the membrane, probably to a great
deal consist of fulvic acids with strong ability to complex metal
ions. One way to minimize the organically complexed forms is to keep
the dialysis time as short as possible.

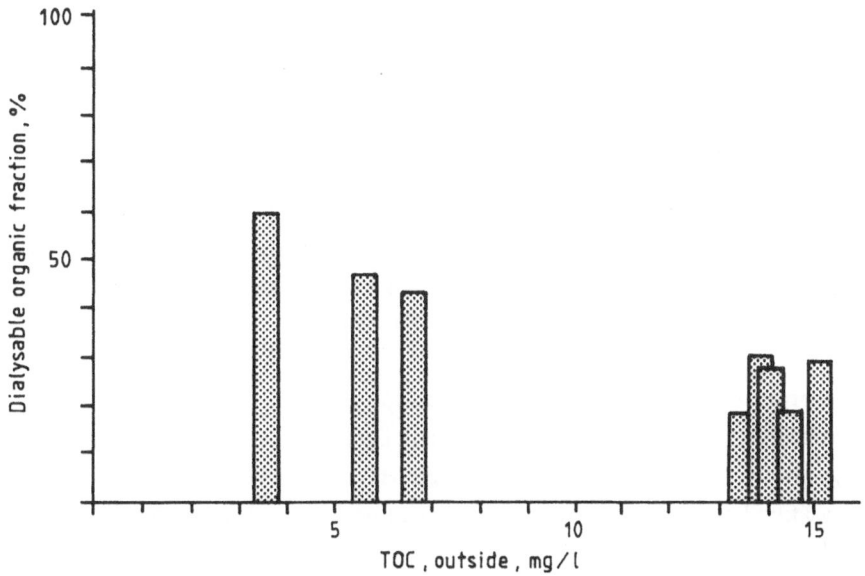

Fig. 4. Penetration of organic substances through a 2000 mwco
membrane. In situ equilibration for 7 days.

Cellulose dialysis membranes provide a good substrate for epiphytic
growth. Experiments in marine environments have shown that at summer
conditions (+20°C), a considerable bacterial population develops

after 3 days and that the membrane is degraded and collapses after 9 days. At winter (+4°C) however, the tubes remain strong and transparent for at least the first 9 days after which biota develops, which after 17 days reach the same density as after 5-7 days in the summer (14). The dialysis experiments discussed in this report were generally performed when the water temperature was 10°C at the most and the dialysis time was limited to 5 days. It is not likely that epiphytic growth have had any substantial influence on the transport of ions through the membrane.

In spite of the potential sources of error discussed above it is still possible to use in situ dialysis for succesful fractionation of dissolved metal species in natural waters, on the condition that attention is given to the possible errors when interpreting the results.

Preliminary results of comparisons between dialysis and cation-exchange separation have shown that in lake waters with low concentration of humus there was a good correlation between the dialysable and the labile inorganic Al-fraction. In more humic waters, the concentration of dialysable Al was higher than the labile inorganic Al, which would be expected as some organic complexes probably penetrate the membrane.

Determinations of the complexing capacity of the waters by potentiometric titration with Cu-solution and Cu-ion selective electrode, showed that generally more dialysable metals were found in waters with lower complexing capacity, i.e. lower TOC-levels (10).

The findings discussed above point to the condition that the dialysable metal fraction is more closely connected to the bioavailable fraction than is the filterable (0.4 μm) fraction or the total concentration. A good correlation between dialysable Cd and biological uptake in mussels have been demonstrated. Cu showed

another accumulation pattern than Cd with a certain treshold
concentration (around 0.2 µg/l) where the mussels started to
accumulate (15).
The _in situ_ dialysis experiments have demonstrated that, in general,
Zn, Cd, As and Mn occur to a high degree in dissolved form in fresh
waters, Cu in dissolved and colloidal form, while Pb, Al and Fe
occur mainly in the particulate fraction except in very acid lakes
with low TOC-concentration. The results indicate that Cd, Zn, and As
generally would be more bioavailable in fresh waters than Cu, Al and
especially Pb.

References

1. Borg, H. Hydrobiologia 101:27-34 (1983

2. Borg, H. Water Res. (in press)

3. Beneš, P. and E. Steinnes. Water Res. 8:947-953 (1974)

4. Beneš, P. Water res. 14:511-513 (1980)

5. Hart, B.T. and S.H.R. Davies. Aust. J. Mar. Freshw. Res. 32:
    175-189 (1981)

6. Borg, H. and P. Andersson. Verh. Internat. Verein. Limnol.
    22: 725-729 (1984)

7. Truitt, R.E. and J.H. Weber. Environ. Sci. Technol. 15: 1204-
    1208 (1981)

8. Morrison, G.M.P. Metal Speciation in Urban Runoff. Ph.D.
    Thesis. Middlesex Polytechnic U.K. 316 pp (1985)

9. Powers, C.D., R.G. Rowland and C.F. Wurster. Water Res. 10:
    991-994 (1976)

10. Borg, H. Metal species in lake waters in the Rönnskär area.
    The National Environmental Protection Board, report 3124, 16
    pp (in Swedish) (1986)

11. Borg, H., I. Gustavsson, G. Johansson and M. Bengtsson.
    Recommendations for sampling and analysis of trace metals in
    natural waters. The National Environmental Protection
    Board,report 1918, 40 pp (in Swedish) (1984)

12.   Borg, H. Water, Air & Soil Pollut. (in press)
13.   LaZerte, B.D. Can. J. Fish. Aquat. Sci. <u>41</u>: 766-776 (1984)
14.   Vargo, G.A., P.E. Hargraves and P. Johnson. Marine Biology
       31: 113-120 (1975)
15.   del Castilho, P., R.G. Gerritse, J.M. Marguenie and W.
       Salomons. In: C.J.M. Kramer and J.C. Duinker, Eds.,
       Complexation of Trace Metals in Natural Waters. Nijhoff/Junk
       Publ. The Hague (1984) pp 445-448.

TRACE ELEMENT SPECIATION IN NATURAL WATERS USING HOLLOW-FIBER ULTRAFILTRATION

E. Lydersen, H.E. Bjørnstad, B. Salbu, A.C. Pappas
Department of Chemistry, University of Oslo
P.O.B. 1033, Blindern, 0315 Oslo 3, Norway

ABSTRACT

In natural waters trace elements may be present in different physico/chemical forms, varying in size, charge and density.

In order to obtain information on the size distribution pattern, hollow-fiber ultrafiltration technique is a useful tool. The membranes are made of inert polymers with different nominal molecular weight cut-off levels. Compared with traditional disc filtering techniques, the main advantages with hollow-fibers are basically the high filtering capacity, combined with minimal clogging and sorption problems. Using a peristaltic pump, the water is transported directly into the molecular weight discriminators. Thus fractionation can be performed in the field, "in situ" fractionation. The system is closed, contamination risks are minimized and it is possible to determine the degree of sorption on the internal equipment surfaces using a mass-balance approach. Applications of hollow-fibers in natural waters and in laboratory studies are demonstrated in this paper.

INTRODUCTION

In general, total concentrations of trace elements give limited information on interactions with biota, and/or the degree of biological availability. In order to achieve such information, samples must be fractionated with respect to chemical and/or physical properties of actual elements. The hollow fiber ultrafiltration technique is a mechanical/physical separation technique where species smaller than the pore-diameter of the membrane are isolated. The fractionation takes place without addition of any reagents.

EQUIPMENT DESCRIPTION

A hollow fiber cartridge consists of several fine cylindrical tubes, each with a lumen diameter of 0.2 or 0.5 mm. The matrix is inert, non-ionic polymers. The high axial flow through hollow fiber lumens produces high shear forces at the membrane surface, minimizing concentration polarization, clogging and sorption by rejected solutes. As water flows through the fiber, it causes micro solutes and salts to penetrate the membrane, while rejected solutes are discharged as waste (figure 1). The filtering pressure is generally about 10-15 psi (69-172 kPa).

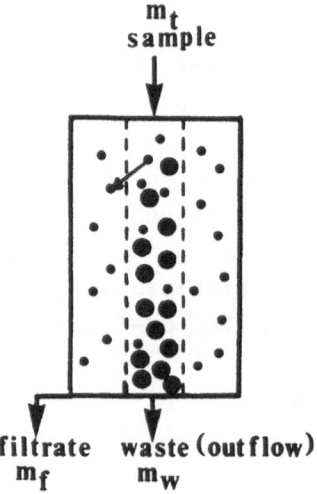

Figure 1: The principal of hollow fiber fractionation

Different hollow fiber membranes are comercially available, having different nominal molecular weight cut-offs in the range $3x10^3$ to 10^6 Dalton. In our laboratory we have systematically used membranes of all cut-offs.

In Table 1 some cartridge specification are listed (1). Flow-rates (ml/min) for deionized distilled water listed in this table are not valid under field conditions as the viscosity increases for low temperature waters, which reduces the filtering flow-rate dramatically, especially for the 10^3 and 10^4 Dalton fibers. High turbidity waters, either due to prescence of organic and/or inorganic components, also affect the filtering flow-rate, but to a lesser extent than the temperature. For 10^6 and 10^5 Dalton fibers the flow-rates are more affected by turbidity than for the 10^4 and 10^3 Dalton fibers.

Cartridge Type	Nominal Cut-off Dalton	Surface Area m^2 ft^2		Number of Fibers	Deionized Water Flow-Rate ml/min[••]	Recommended Recirculation Rate l/min
H1P1 - 43	1,000[•]	.03	.3	55	20-30	0.6-1.8
H1P10 - 8	10,000[•]	.09	.9	1000	170-320	0.2-0.6
H1P10 - 20	10,000	.06	.6	250	60-90	0.3-0.9
H1P100 - 20	100,000	.06	.6	250	100-140	0.3-0.9
H1MP01-43	0.1 µm	.03	.3	55	800-1100	0.6-1.8

[•] : not available
[••] : transmembrane pressure of 10 psi (69 kPa) at room temp. (ca. 20°)

Table 1:   Some hollow fiber cartridge specifications according to the manufacturer (1).

WASHING PROCEDURE

Before and after use, the hollow fiber cartridges have to undergo a thorough washing procedure. In Figure 2 the cleaning solutions used in the presented work are listed which are in accordance to the recommendation given by the manufacturer (1). These solutions should be prepared with high purity water (e.g. deionized distilled water). In addition, a chelating agent, like EDTA, should also be included in the procedure, in order to desorbe metals from internal surfaces. After cleaning, the cartridges should be rinsed with high purity water, at least 15-20 l.

MEMBRANE CUT-OFF MEASUREMENTS

In order to achieve information about pore-size distributions of the membranes, standard solutions containing chemically well defined species e.g. globular proteins, carbohydrates can be used (2,3).

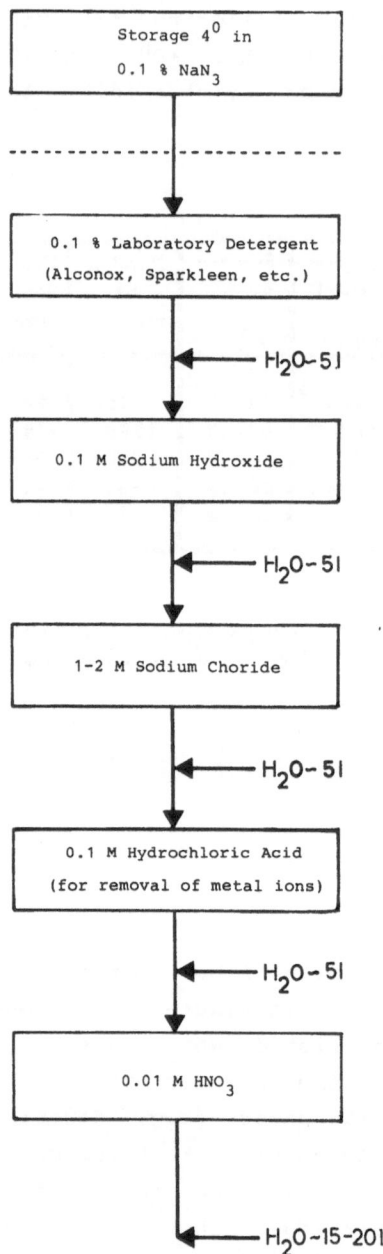

Figure 2: A recommended hollow fiber washing procedure

Figure 3 illustrates the rejections of different compounds by a 10^4 Dalton membrane (H1P10-8)(4). A steep curve indicates a narrow pore-size distribution.

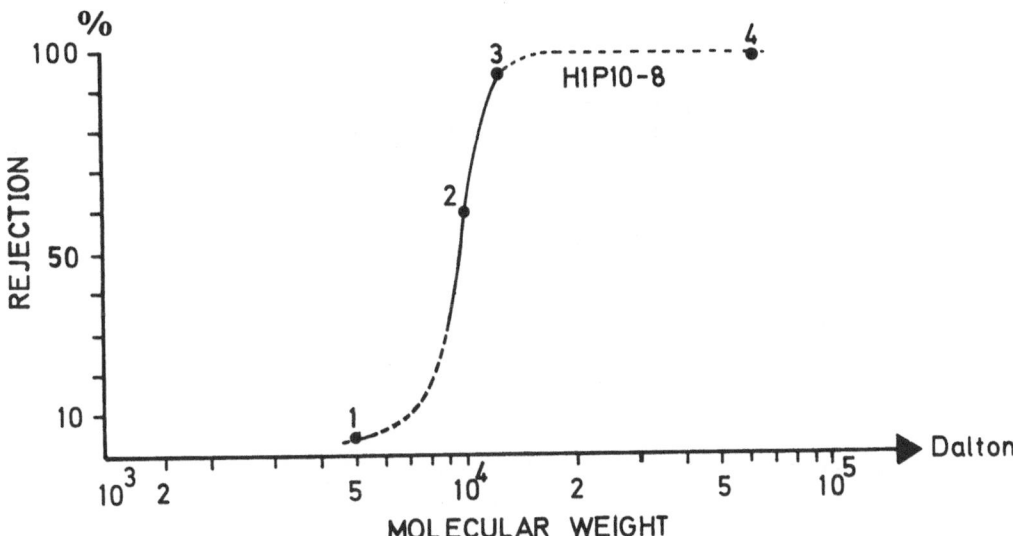

Figure 3: Rejection characteristics for hollow fiber H1P10-8
(cut-off:10^4 Dalton), using different solutes with varying
nominal molecular weight average. 1 : Inulin = 5 000 mw, 2
: PVPK-15 = 10 000 mw, 3 : Cyt.Crom C = 12 500 mw and 4 :
Albumin = 67 000 mw (4).

CONTAMINATION - CLOGGING - SORPTION PROBLEMS

Working with trace concentrations, contamination problems from both
the external and internal milieu may seriously affect the results. "In
situ" fractionation in a closed system minimizes external
contamination risks (Figure 4). Furthermore the washing procedure
(Figure 2), followed by a thorough conditioning with the sample itself
(10-15 1), internal contamination risks are minimized too. One main
advantage using the hollow fiber fractionation technique, is that
sorption and clogging can to a certain degree be controlled. From a
mass-balance calculation, an operation-quality factor can be defined:

$$p = \frac{m_f + m_w}{m_t} \qquad (i)$$

where $\qquad$ m_t = total mass in inflowing water

$\qquad\qquad\qquad$ m_f = total mass in filtrate water

$\qquad\qquad\qquad$ m_w = total mass in outflowing water (waste)

"IN SITU,,-fractionation

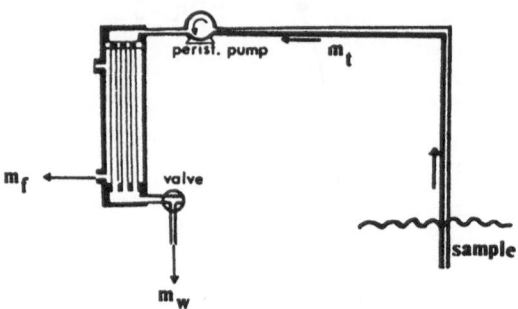

Figure 4: Schematic set-up of a hollow fiber system under "in situ"
fractionation sampling (5).

If p < 1, there is a loss, either due to sorption into internal
surfaces and/or clogging of the membrane.

If p > 1, the water is contaminated by internal sources.

If p = 1, the filtration is probably not affected by these effects.

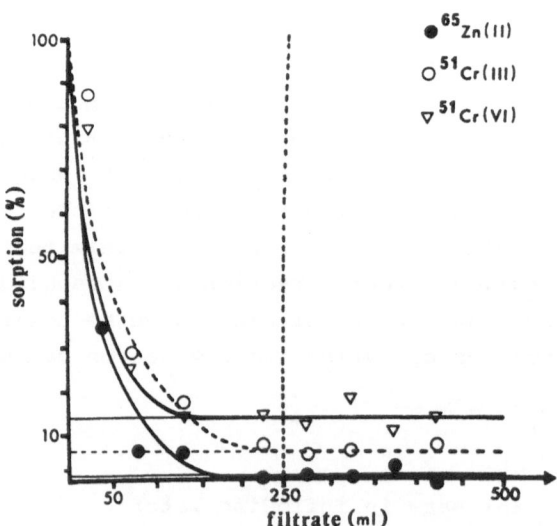

Figure 5: Sorption of Zn (II), Cr (III) and Cr (VI) on a H1P10-8
hollow fiber (cut-off : 10^4 Dalton)(6).

Information about sorption effects in connection with membranes, in speciation studies of trace elements, is often obtained using artificial spikes. Radio-nuclides are very useful in these studies (Figure 5). ^{65}Zn (II) and ^{51}Cr (III) as chlorides and ^{51}Cr (VI) as sodiumcromate were added (0.5 μg/l with spec. activity : ^{65}Zn = 8 MBq/μq Zn, ^{51}Cr = 12 MBq/μg Cr and ^{51}Cr (as cromate) : 14 MBq/μg Cr) to ultrafiltered deionized distilled water (6). As shown in Figure 5, after about 250 ml filtrate, the sorption is minimized to about 0, 7 and 13% respectively, indicating that the samples can be collected. For the 10^4 Dalton cartridge (H1P10-20) that means a total conditioning water consume of about 10-15 l. However, in natural water samples, there will be inorganic and organic species other than used above, that may be preferentially sorbed. Thus clogging and sorption effects were studied in natural waters. Water samples from lake Nepptjern and lake Lomstjern, two small acid lakes near Oslo, Norway, were studied (Table 2)(5). These lakes were chosen because of their approximately same levels of total monomeric aluminium (Al$_a$) and pH. The main difference is the total organic carbon (TOC) content (Table 2).

	NEPPTJERN	LOMSTJERN
$\varkappa_{20^\circ}$	28 μS/cm	30 μS/cm
pH	5.0	5.1
Al$_a$*	596 μg/l	518 μg/l
TOC	1.7 mg/l	13.7 mg/l

* Total monomeric aluminium

Table 2: Some physical/chemical parameters from lake Nepptjern and lake Lomstjern (5).

According to the mass-balance calculation (i), sorption/clogging losses may be estimated.

$$m_s\% = (1 - p) \times 100 \qquad (ii)$$

where $m_s\%$ = Internal mass loss/gain (%) due to sorption/clogging and contamination

Figure 6 is based on these calculations. In this work the determination of total monomeric aluminium (Al_a) is based on the Barnes/Driscoll method (7,8).

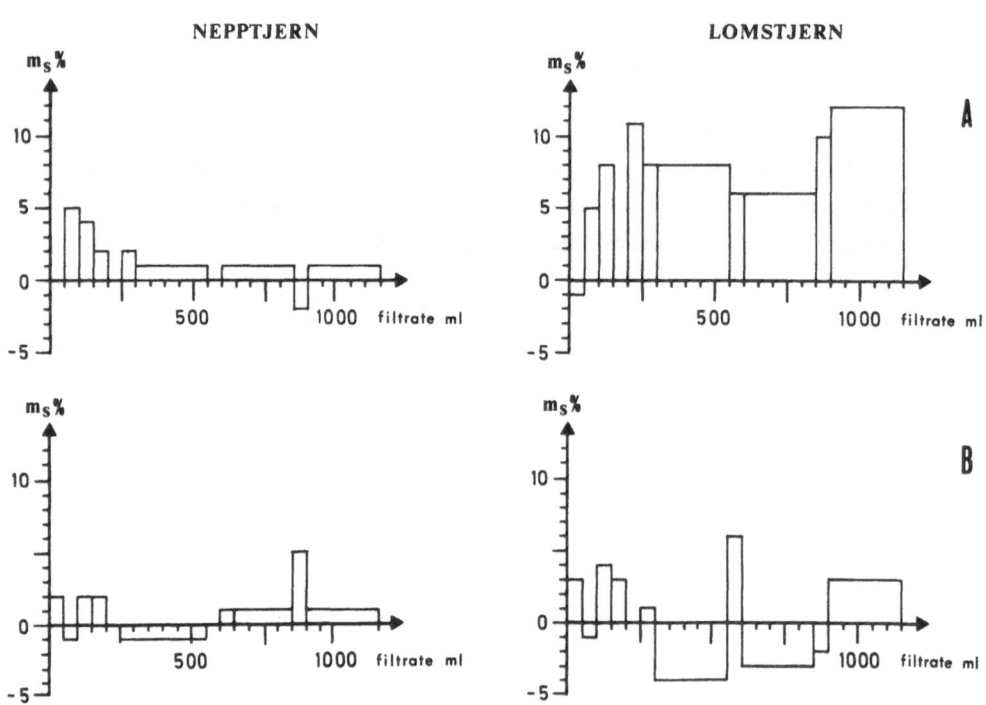

Figure 6: Mass sorbed (m_s%) of total monomeric aluminium (Al_a) under 10^4 and 10^5 Dalton filtration, based on the mass-balance calculation (ii), in lake Nepptjern and lake Lomstjern. A : 10^5 Dalton fiber. B : 10^4 Dalton fiber.

For both fibers, 10^4 and 10^5 Dalton (H1P10-20, H1P100-20), the mass-balance calculations gave reasonable results. In the 10^5 Dalton fiber a mean net loss (positive value) of about (1 ± 1)% was observed in the low TOC lake, and (8 ± 4)% in the high TOC lake. The difference in net loss in these lakes is small, but significant. The higher retention (m_s%) in the high TOC lake is probably due to the higher TOC content, which seems to cause a small sorption/clogging effect. Using 10^4 Dalton fibers, it is difficult to discover sorption/clogging effects in the high TOC lake compared to the low TOC lake. In the 10^4 Dalton fiber mean net loss was about (1 ± 1)% and (0 ± 3)% in the low and high TOC lake, respectively. Therefore the sorption/clogging losses of Al_a seem to be small and scarcely significant.

The total flow-rate through the fibers is the same for all fibers (ca. 300 ml/min) due to the peristaltic pump-speed. The main reason for reduced sorption/clogging in the 10^4 Dalton fibers is due to a higher waste water flow-rate, and a corresponding reduction in the flow-rate of the filtrate.

During 10^4 Dalton fiber filtrations the waste water flow-rate is high, leading to an important washing effect of the tube interior. This washing effect minimizes sorption/clogging, and mostly affects the 10^3 and 10^4 Dalton fibers. Due to lower waste water flow-rate, but higher flow-rate of filtrate in the 10^5 and 10^6 Dalton fibers, a reduced washing effect leads to higher sorption/clogging. These results indicate that sorption/clogging effects seem to be minimal with hollow fiber fractionations compared to traditional filtering techniques.

HOLLOW FIBER FRACTIONATION - SOME RESULTS

The hollow fiber fractionation was carried out in a high mountain catchment, Åstadalen, 200 kilometers north-northeast of Oslo, Norway (12). Some results on distribution patterns for aluminium, iron and zinc from different watersheds, A-E (Figure 7)(12), are presented here. Two incident studies have been carried out, one at high flow, during a heavy rain period, and one at low flow, during a summer stagnation period.

The pH, as an average for the whole catchment, was about 6 ± 1 at low flow, but during high flow a marked pH-drop of about 0.5 to 1 pH-unit was observed.

As illustrated in Figure 7, a substantial fraction of aluminium, iron and zinc is associated with colloids, pseudo-colloids and particles, i.e. mw > 10^3 Dalton. The relative distribution patterns depend on the origin of water, seasonal variations etc. At all sites, except for the spring site (A), the total element concentrations increased at high flow. This spring site (A) originates from a small ground water pool and the decrease in total element concentrations at high flow, is a dilution effect probably due to heavy rainfall (12).

The most interesting results from this hollow fiber fractionation studies, is the observed changes in molecular weight distribution. On the whole, there is a lot more low molecular weight species observed at high flow than at low flow. This underlines the importance of doing fractionation studies.

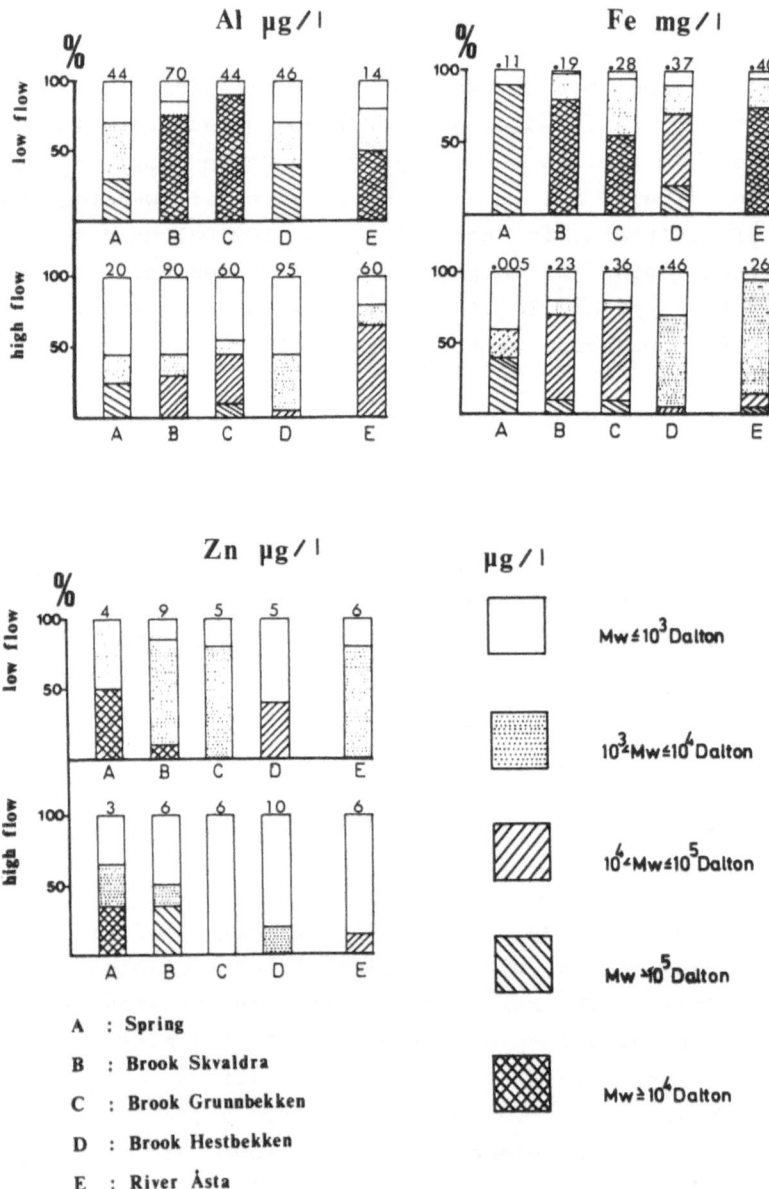

A : Spring
B : Brook Skvaldra
C : Brook Grunnbekken
D : Brook Hestbekken
E : River Åsta

Figure 7: Distribution of Fe, Al and Zn (% of total concentrations) according to size (hollow fiber filtration) at different sites in the Åstadalen catchment (12).

OTHER APPLICATIONS

In addition to "in situ" fractionation, the hollow fiber system has
many other applications, as for instance association studies. In such
studies a continuous diafiltration unit is used, as illustrated in
figure 8 (9). This system has one mixing and one test chamber, and is
kept going by a peristaltic pump. The solution in the mixing chamber
contains high molecular compounds, e.g. colloids etc., completely
rejected by the membrane. The solution in the test chamber contains
radio-nuclides, metal ions etc. These low molecular weight species
enter the mixing chamber and can then associate with colloids.
Observing the decrease in the concentration of low molecular weight
species in the test chamber, it is possible to estimate some
complexing rates and quantify amounts complexed, as illustrated in the
references 9 and 13.

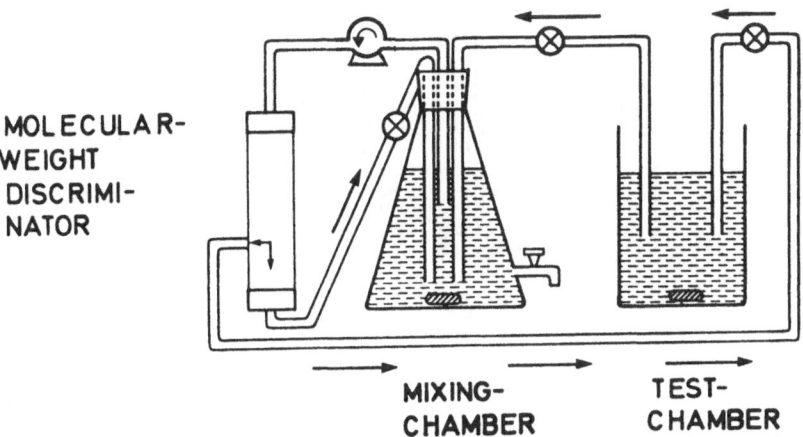

Figure 8: Schematic set-up of a hollow fiber system under association
studies (9).

The hollow fiber system can also be used in diffusion/dialysis studies
where both diffusive and convective transport take place across the
membrane (Figure 9)(10). One chamber contains high purity distilled
water, and one a solution of compounds of unknown molecular weights.
Molecular weight is the far most important parameter as far as
dialysis rates are concerned (11). After having standardized known
molecular weight standards and their dialysis rates, the molecular
weight of unknown species may be estimated (11).

CONCLUSIONS

The main advantages with hollow fiber fractionation are basically the high filtering capacity combined with minimal clogging and sorption problems, compared to traditional disc filtering techniques. The equipment is easy to handle and the whole fractionation can be carried out in a closed system in the field. Based on mass-balance calculations, internal contamination, sorption/clogging effects can be controlled (calculations (i), (ii)). It is also very easy to perform association-, diffusion/dialysis-studies with stable elements and radio-active nuclides.

a) SAMPEL CHAMER
b) MOLECULARWEIGHT DISCRIMINATOR
c) DIALYZATE CHAMBER

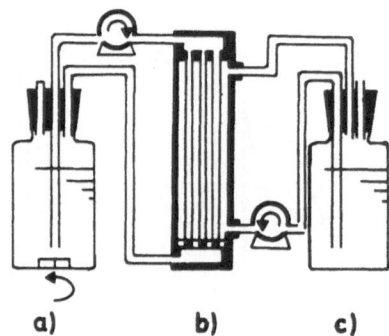

a)   b)   c)

Figure 9:   Schematic set-up of a hollow fiber system under diffusion/dialysis studies (10).

ACKNOWLEDGEMENT

The authors thank the Norwegian Hydrological Committee, the Surface Water Acidification Project (SWAP) and the Norwegian Research Council for Science and the Humanities for financial supports.

(1)   Amicon Corporation. Pub. No. I-IIIB, Amicon Corp., Mass., U.S.A. (1983).

(2)   Ogura, N. Marine Biology 24 : 305-312 (1974).

(3)   Buffle, J., Delaoey, P. and Haerdi, W. Anal. Chim. Acta 24 : 339-357 (1978).

(4)   Salbu, B. In Doctoral thesis. Preconcentration and Fractionation Techniques in the Determination of trace Elements in Natural Waters - their concentration and physico-chemical forms. University of Oslo, 106-114 (1984).

(5)   Bjornstad, H.E., Lydersen, E., Salbu, B., Sullivan, T., Munitz, I.P. and Voght, R. Size Fractionation of Al-species in Acidified Fresh Waters. (In prep.).

(6)   Salbu, B., Bjornstad, H.E., Lindstrom, N.S. and Lydersen, E. Talanta 9 : 907-913 (1985).

(7)   Barnes, R.B. Chem. Geol. 15 : 177-191 (1975).

(8)   Driscoll, C.T. Intern. J. Environ. Anal. Chem. 16 : 267-283 (1984).

(9)   Bjornstad, H.E. and Salbu, B. Talanta (1986, in press).

(10)   Salbu, B., Bjornstad, H.E. and Pappas, A.C. In : R.A. Bulman and J.R. Copper, Eds., Speciation of Fission and Activation Products in the Environment. Elsevier Appl. Scient. Publ., London, New York, p. 101 (1985).

(11)   Craig, L.C., King, T.P. and Stracher, A. J. Am. Chem. Soc. 79 3729-3737 (1957).

(12)   Salbu, B., Bjornstad, H.E., Bibow, J.O., Englund, H., Hovin, H. and Rambaek, J.P. Trace Elements in Fresh Waters from Åstadalen, a High Mountain Catchmentin S.E. Norway. (In prep.).

(13)   Salbu, B., Bjornstad, H.E., Lydersen, E. and Pappas, A.C. Determination of Radionuclides Associated with Colloids in Natural Waters. In : Proc. Int. Conf. Low Level Measurements of Actinides and Long-Lived Radionuclides in Biological and Environmental Samples, Lund, June 9-13 (1986, in press).

THE IMPORTANCE OF SORPTION PHENOMENA IN RELATION TO TRACE ELEMENT SPECIATION AND MOBILITY

B. Allard, K. Håkansson and S. Karlsson
Department of Water in Environment and Society
Linköping University, S-581 83 Linköping, Sweden

INTRODUCTION

The transport of water in environmental systems, e.g. surface waters and groundwaters, can in principle be described in terms of advection and dispersion phenomena as well as diffusion. Constituents in the water, either in true solution or as suspended matter, can undergo chemical transformations or interact with solid phases present in the system and will therefore not generally be described by the same transport models as the water itself. Dissolved trace elments in particular are prone to participate in sorption processes and will in most cases be retarded in relation to the physical transport of the non-reacting water. Diffusion of trace components into micro fissures in water-exposed solid materials would in fact lead to a retention even in the absence of any specific sorption reaction.

The importance of sorption phenomena for the transport of dissolved micro components in e.g. environmental waters and the relation of these processes to chemi cal speciation are reviewed in this paper. Examples are given from a case study of metal releases from a mine waste deposit.

SORPTION PHENOMENA

The removal of dissolved trace elements from an aqueous phase is the result of several processes, notably adsorption reactions or precipitation. There are, however, no distinct borders between the various types of reactions, and the mechanisms for different sorption processes as well as precipitation/coprecipitation etc. can rarely be unambiguously characterized.

Adsorption processes

In general some idealized sorption mechanisms can be distinguished, as described below [1-4].

Physical adsorption is due to non-specific forces of attraction (van der Waals forces) involving the entire electron shells of the dissolved trace element and of the adsorbent. The process is rapid, reversible and not particularly dependent of temperature and the ionic strength of the solution (well below saturation of any solid trace element phase). The presence of complexing agents as well as pH of the aqueuous phase have a large influence on the process.

Electrostatic adsorption (ion exchange) is due to coulombic forces of attraction between charged solute species and the adsorbing phase. The process is usually rapid, largely reversible, somewhat dependent on temperature and strongly dependent on the composition of the solution as well as of the sorbent.

Specific adsorption (chemisorption) is due to the action of chemical forces of attraction leading to surface bonds to specific sites on the solid phase. The process can be slow and partly irreversible. It is highly dependent on the composition of the solid surface as well as of the concentration of the solute, pH and temperature.

Chemical substitution (coprecipitation or solid solution) can be compared with adsorption reactions in the sense that the result of these processes would be a removal of a trace constituent from the solution phase.

The various types of sorption are influenced by physical and chemical parameters of the system such as
* the hydrogen ion availability of the aqueous phase (defined by pH)
* the presence of complexing agents in the aqueous phase
* the free electron availability of the aqueuous phase (defined by Eh)
* the trace element concentration
* the ionic strength of the aqueous phase
* the composition and surface properties (e.g. charge) of the solid phase

Thus, the sorption of a particular trace element on a solid surface is largely dependent on its chemical state, which is illustrated in Figure 1. In principle, the interaction of every species in solution with every single different component of the solid phase can be described in terms of a chemical reaction with a formation constant.

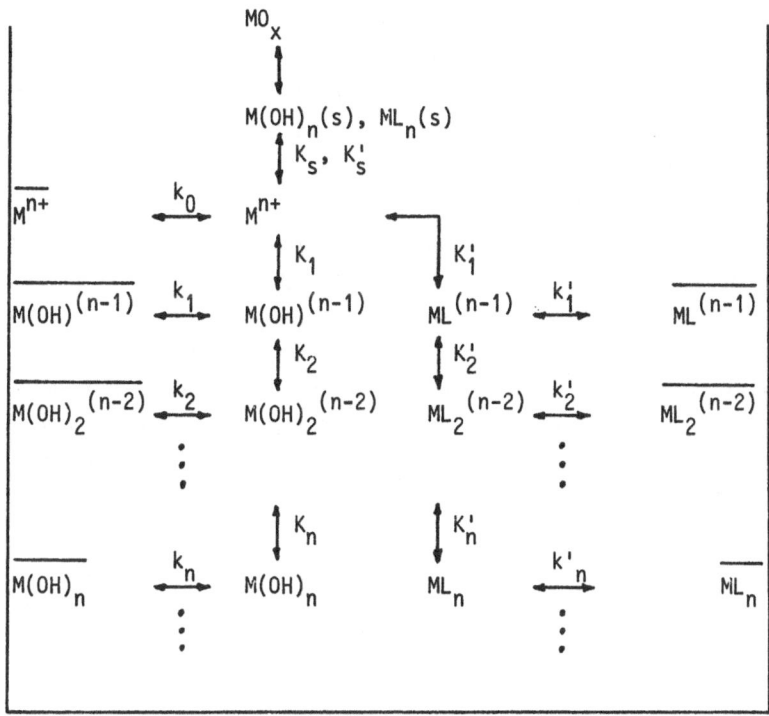

<u>Figure 1</u> Simplified schematic of speciation and sorption of metals in solution
considering hydrolysis and the presence of a complexing ligand L (adsorbed species denoted by a bar)

The interaction between solute and sorbent (sorption-desorption) can either be
described as an equilibrium situation or as a kinetic process. The interaction
with inner surfaces (in micro fissures and between micro crystals etc.) in the
solid sorbent phase would be diffusion controlled. Therefore, the sorption of
dissolved matter on solids would often exhibit a fast initial phase (corresponding to the interaction with easily accessible outer surfaces), followed by a slow
and partly irreversible process (corresponding to sorption within the solid matrix). Thus, true equilibrium would require long reaction times, unless the solid
phase consists of very small particles with short distances for diffusion. A
truly reversible reaction (desorption) can in principle never be expected, since
diffusion into the matrix and sorption would continue even under conditions where
desorption would prevail on the outer surfaces of the solid sorbent phase.

The distribution coefficient concept

The sorption of a dissolved element on a solid can easily be experimentally determined and described e.g. in terms of concentrations in the two phases (solid and aqueous phases) under the particular condition of the measurements. The distribution coefficient (Kd) is defined as the ratio of concentrations, i.e. mass or activity of the element which is bound to the solids present divided by the corresponding amount remaining in the solution phase (units $L^3 M^{-1}$ or $L^3 L^{-2}$). This is a convenient way of expressing the distribution of the element between the two phases, e.g. from analyses under controlled conditions or from field observationsIt should be recognized that Kd is a purely experimental parameter which is highly dependent on the chemical and physical properties of the system and only defines the over-all distribution of the element regardless of chemical state or non-equlibrium conditions.

Various sorption models utilizing more sophisticated expressions defining the sorption of a trace element has successfully been applied in simple well defined systems (isotherm models, mass action models, statistical models, double layer or surface complexation models, etc.).

Sorption behaviour of Cu, Zn, Cd and Pb

The sorption of hydrolysable transition elements, e.g. the divalent metals Cu, Zn, Cd and Pb, onto hydrous oxide surfaces has been described according to a variety of models, e.g.
* the Gouy-Chapman-Stern-Graham model (accounts for specific and electrostatic adsorption),
* the adsorption-hydrolysis model (sorption related to the degree of hydrolysis)
* the ion-solvent interaction model (considers coulombic, solvation and specific chemical interactions)
* the ion exchange model (cations replacing protons)
* the surface complexation model (hydrous oxide surface groups treated as complex forming species).

Also differences between the bulk solution and the surface or aqueous surface film of the solid could lead to:
* precipitation/coprecipitation (redox changes, formation of sparingly soluble surface compounds etc.).

Regardless of the choice of model for the theoretical evaluation of the observed sorption behaviour the distribution of hydrolyzable metals between hydrous oxides and aqueous solutions is very strongly related to pH of the aqueous phase, as illustrated in Figure 2.

Data in Figure 2 origin from laboratory experiments under well-defined controlled conditions (a single sorbent, constant solid/solution ratio; control of trace element concentration and pH in an aqueous phase with a constant ionic strength; controlled contact times, etc.). Under field conditions there will be variations with time and location of important parameters such as

* composition of the aqueous phase (presence of complexing agents, pH, salt content; mixing of surface waters, groundwater and precipitation)
* compositon of solid phases (presence of a heterogeneous stationary solid phase as well as of suspended matter of various composition and particle sizes)
* hold-up times and flow conditions (various degree of equilibrium; kinetic effects due to mixing conditions, flow conditions, diffusion controlled sorption processes, precipitation, formation of large aggregates/coagulation as well as formation of colloidal phases).

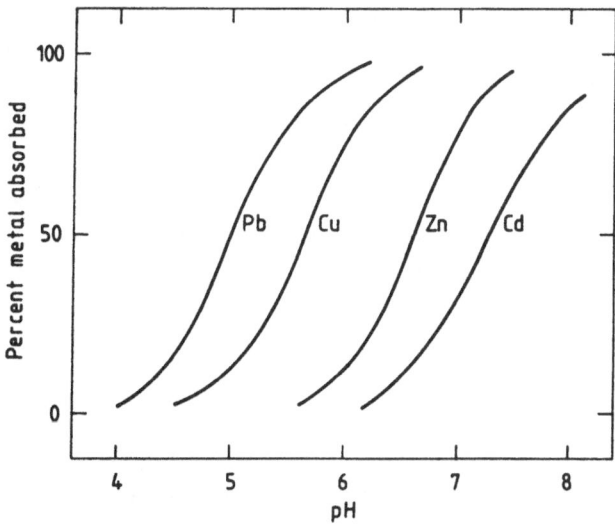

Figure 2 Adsorption of Cu, Zn, Cd and Pb on amorphous $Fe_2O_3 \cdot H_2O$ [5] (Total metal concentrations 5 x 10^{-5} M).

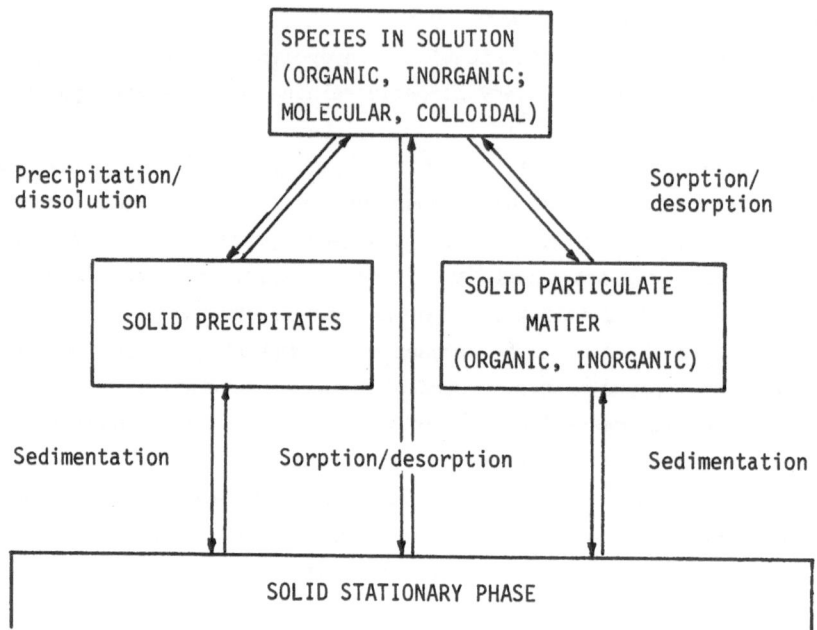

<u>Figure 3</u> Distribution of a trace element between solid phases and solution phase

The complicated multi-component multi-phase system encountered in nature is illustrated in Figure 3. Thus, the difficulties in explaining or predicting the quantitative distribution and transport properties of trace metals under environmental conditions are substantial, despite the large volume of data from sorption studies under ideal laboratory conditions.

FIELD OBSERVATIONS

The data presented below originate from field observations of the effluents from a sulfide mine waste deposit at Bersbo, some 250 km SSW of Stockholm in Sweden.

<u>The Bersbo area</u>

Copper was mined at Bersbo as early as in the 14th century and continued until the early 20th century. A period of rational mining from 1765 to 1902, lead to a peak production during a few decades around 1860-1870. The host rock in the area

is granite, while the principal ore is chalcopyrite in association with pyrite.
The pyrite also contains minor quantities of sphaelerite and galena. Tailings
from the mining, mostly from the late 19th century, cover an area of approximately 0.2 km^2 and consist of waste materials with grain sizes from silt to rock
(totally some 300 000 m^3 in the area). The deposit area and one of the major
drainages of leachates are illustrated in Figure 4.

Hydraulic conductivities within the deposit have been estimated to be in the
range of 10^{-3} m s^{-1}. Thus, the material is readily exposed to precipitation.
Weathering processes lead to an oxidation of the sulfides, generating an acidic
leachate (ph ca 3) with high concentrations of sulfate (more than 1 g l^{-1}) and
dissolved transition metals (particularly Fe, Mn, Zn, Cu, Cd, Pb; e.g. more than
0.1 g l^{-1} of zinc).

Water has been sampled weekly and analyzed (pH, major anions and cations; Fe,
Cu, Zn, Cd and Pb from the leachate) since early 1983. Detailed anlyses of the
observed metal mobilities in the field are in progress. Efforts are made to model
the quantitative out-flow of dissolved matter fom the deposit as well as the pH-
changes in the down-stream river and lake system [6-11]. The data given below are
parts of this study, selected to serve as a demonstration of the various processes indicated in Figure 3.

Metal adsorption on solid phases

The unpolluted water (location 1, Figure 4) generally has a pH of 4.5-6 and
low alkalinity (total carbonate < 10 mg l^{-1}). Total salt content is in the range
50-100 mg l^{-1} inorganic constituents, and organic material is usually in the
range 5-20 mg l^{-1}. The dominating anions are sulfate (25-50 mg l^{-1}) and chloride
(5-15 mg l^{-1}). The principal cations are calcium (5-12 mg l^{-1}), sodium (5-15 mg
l^{-1}), magnesium (5-10 mg l^{-1}) and potassium (1-5 mg l^{-1}).

The in-flow of acidic leachates causes a drastic change in the composition,
particularly for pH and sulfate (3-3.5 and < 1500 mg l^{-1}, respectively in the
leachate). Down-stream, pH is gradually decreasing, due to mixing with unpolluted
water as well as neutralization reactions. The sulfate concentration is gradually
decreasing, seemingly as a function of pH according to [10]:

$$\log(SO_4{}^{2-}) = -0.3 \text{ pH} - 1.2 \text{ (M)}$$

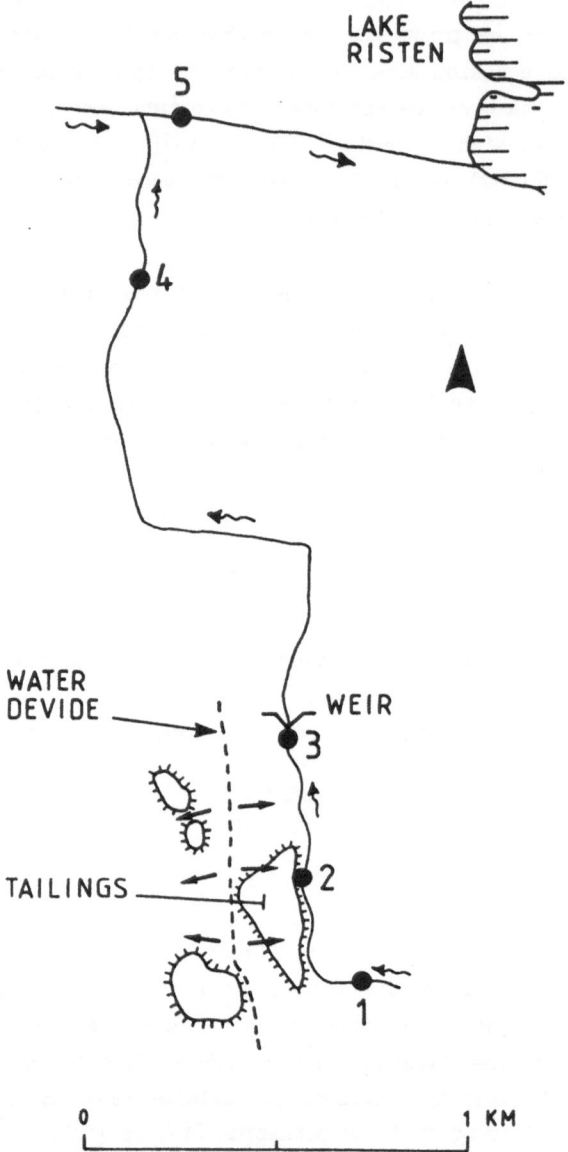

Figure 4   The Bersbo area [6]

(Numbers indicate water sampling locations)

Carbonate concentrations are increasing, roughly determined by the equilibrium with atmospheric carbondioxide:

$$\log (CO_3^{2-}) = 2\ pH - 19.8 \qquad (M) \qquad (pH\quad 6.5)$$
$$\log (CO_3^{2-}) = pH - 13.3 \qquad (M) \qquad (pH\quad 6.5)$$

The total concentrations of dissolved trace metals (analyses of filtered samples, 0.4 μm polycarbonate filters) as a function of pH are given in Figure 5 (for the period May 1984 - August 1984).

Calculated solubilities (considering pH, SO_4^{2-} and CO_3^{2-}) are significantly above the observed concentrations, except possibly for cadmium. Thus, the decrease in metal concentrations as a function of pH would reflect sorption on

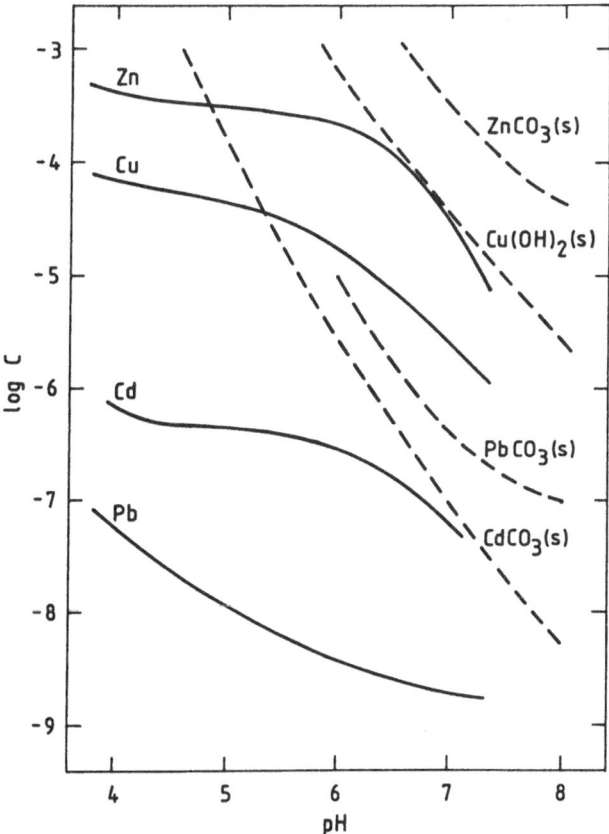

Figure 5 Observed metal concentrations (C, M) as a function of pH [10]
(Dashed lines indicate calculated solubilities for some possible
solid phases).

solids and the dilution down-stream the location of the in-flow of leachate rather than precipitation of sparingly soluble metal compounds.

The dilution due to mixing of polluted and unpolluted water can be assessed from the sulfate balance under the assumption that the sulfate concentration is constant and low in the unpolluted waters in comparison with the leachate. Thus, from the sulfate balance, a dilution factor can be estimated, and the fraction lost due to sorption processes can be calculated [10]:

$$F = (1+x)Q(M_{calc} - M_{obs})M_{obs}^{-1}$$

where x = the dilution factor, Q = water flow, M_{calc} = calculated metal concentration assuming dilution only and M_{obs} = observed metal concentration. The function F is a distribution function describing the sorption onto solid sorbents in the stream (disregarding differences in mixing and contact times for various Q). The sorption of trace metals defined by the function F is illustrated in Figure 6. The observed sorption behaviour is in fair agreement with distribution functions obtained in laboratory studies [5] considering the heterogeneity in flow-conditions, slow kinetics and the approximate estimate of the dilution.

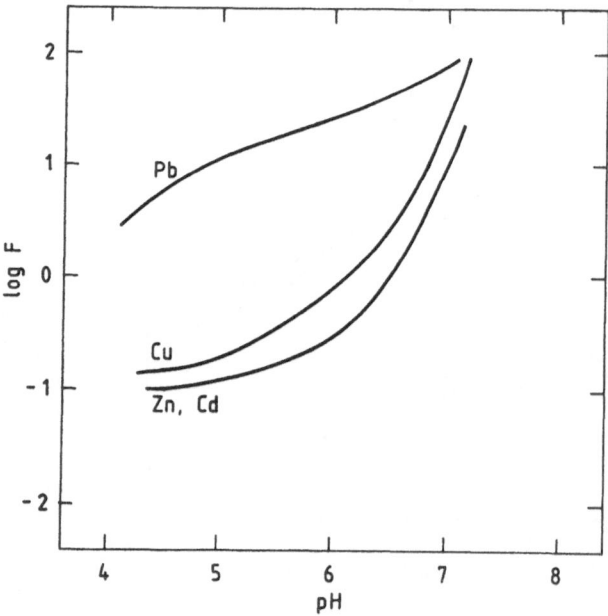

Figure 6 Fraction of trace metals lost due to sorption processes; F as a function of pH [10].

The fraction of the metals that are associated with particulate matter (retained by filtration through 0.40 μm filter) is illustrated in Figure 7 [9].

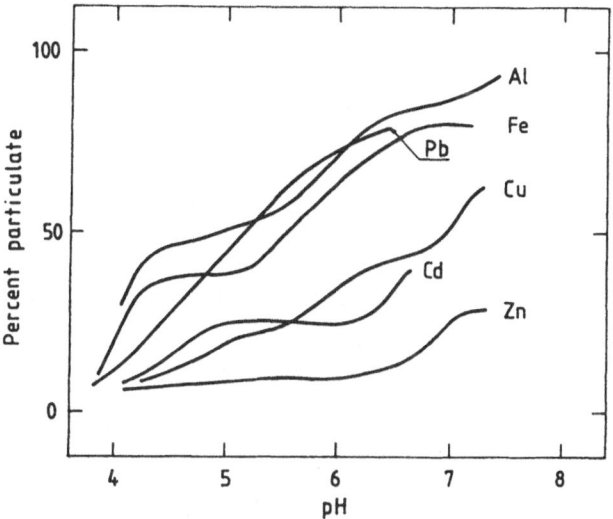

<u>Figure 7</u> Metal distribution between suspended matter ($>$0.40 μm) and solution as a function of pH [9].

Sorption mechanisms

Sorption on the solid suspended phase (or precipitation/coprecipitation) is the major mechanism for the removal of metals from the the aqueous phase. Calculations of theoretical solubilities, assuming the formation of hydroxides, carbonates, sulfates, hydroxycarbonates or hydroxysulfates under the present conditions show that aluminum and iron both reach saturation already at pH around 4 [11]. Possibly, also cadmium could reach saturation at high pH, as well as copper and lead (basic carbonates). The pH-dependence of the removal of Cu, Zn and Cd rather indicates a sorption on the fresh amorphous Al-Fe-hydroxy precipitate and a subsequent coprecipitation and occlusion as the mechanism for the removal.

The suspended solid matter ($>$0.40 μm) contains mainly aluminum and iron (from 20 to almost 100% assuming hydroxides) with some additional silica (less than 5%). The remaining fraction (up to 76%) is organic [11].

The solid is largely amorphous, although minor quantities of crystalline matter (possibly iron hydroxide and basic lead carbonates as well as quartz) are indicated from X-ray powder diffraction data.

The suspended solids were characterized by a sequential leaching procedure [12, 13] identifying the following metal fractions:
* Exchangeable and bound to carbonates and hydroxides
* Bound to (hydrous) oxides
* Bound to organics and sulfides
* Residual.
Some results are given in Table I.

Table I Sequential leaching of suspended solids (pH 6.51, oxygen saturation)

	Exchangable/carbonates	Oxides	Organic	Residual	Total
	%	%	%	%	mg/g
Al	15	19	26	40	32
Fe	23	59	9	9	39
Cd	93	5	1	1	0.010
Cu	52	8	39	1	0.50
Pb	53	38	3	6	0.034
Zn	79	15	3	3	4.0

The exchangable fraction appears to be the most significant one for all elements, with lead also largely present in the organic fraction. Although solid (amorphous?) oxides are precipitating in the stream the heavier elements are only present as coprecipitates to a minor extent.

Organic material in association with the solid surfaces are binding copper and lead. It is possible that this reaction limit the coprecipitation of these elements with the precipitating solid phases.

Sedimentation of suspended solids

The suspended solids are incorporated into the bottom sediments under periods with low water flow. Dissolved species can also contribute to the metal content in the sediments when pH is sufficiently high. Sediment cores were collected at the same location as discussed previously and treated with the same leaching procedure as was applied to the suspended solids. Some leaching data are given in Table II.

Table II Sequential leaching of sediments (0-10 cm; pH 7.0)

	Exchangable/carbonates %	Oxides %	Organic %	Residual %	Total mg/g
Al					0.8
Fe					100
Cd	50%	30%	15%	5%	0.034
Cu	10%	1%	70%	10%	2.4
Pb	30%	10%	10%	45%	0.15
Zn	30%	25%	40%	5%	7.1

The exchangable fraction appears to be decreased in comparison with the suspended solids. Amorphous oxides are to a greater extent contributing to metal binding which might be attributed to variations in redox conditions. The increase in organic and mineral fractions are also liable to depend on transformations of suspended oxides for reasons mentioned above.

DISCUSSION AND CONCLUSIONS

Sorption to precipitating solids can be the most important mechanism in controlling apparent solubility of Cd, Cu and Zn in undersaturated solutions. This process is highly dependent on pH in the system. Lead appears to be readily incorporated into more stable chemical forms either by coprecipitation or formation of solids not included in this discussion.

Elemental distribution between different fractions in the solid sediments are different in comparison with the suspended solids. Elements are typically present in more "stable" associations in the sediments. Redistribution of settling particles from the aqueous phase leads to chemical changes.

The solute speciation appears to be more important for sorption than the composition of the solid itself. This is also apparent in the sediment but in this environment it is also of great importance to consider the stability of solids in relation to important chemical parameters such as pH and redox conditions. The general schematic description in Figure 3 of the distribution of trace elements between the solution phase and solid phases (mobile or stationary) and the corresponding processes determining this distribution (complexation, sorption/desorption, precipitation/dissolution, sedimentation) is well illustrated in the field example.

REFERENCES

1. Benes, P. and V. Majer. Trace Chemistry of Aqueous Solutions (Amsterdam: Elsevier, 1980).

2. Förstner, U. and W. Salomons. In: G.C. Leppard. Ed., Trace Element Speciation in Surface Waters and its Ecological Implications, pp. 245-273. (Burlington: National Water Research Inst., 1983).

3. Muller, A.B., Ed., Sorption. Modelling and Measurement for Nuclear Waste Disposal Studies. (Paris: OECD/NEA, 1983).

4. Solomons, W. and U. Förstner. Metals in the Hydrocycle. (Berlin: Springer, 1984).

5. Benjamin, M.M. and J.O. Leckie. J. Colloid Interface Sci. $\underline{79}$: 209-211 (1981).

6. Allard, B., S. Bergström, M. Brandt, S. Karlsson, U. Lohm and P. Sandén. Nordic Hydrology (1987, in press).

7. Brandt, M., Bergström, S. and P. Sandén. Nordic Hydrology (1987, in press).

8. Sandén, P., S. Karlsson and U. Lohm. Nordic Hydrology (1987, in press).

9. Karlsson, S., P. Sandén and B. Allard. Nordic Hydrology (1987, in press).

10. Sandén, P., S. Karlsson and B. Allard. Water Res. (1987, in press).

11. Karlsson, S., B. Allard and K. Håkansson. Geochim. Cosmochim. Acta (1987, in press).

12. Tessier, A., P.G.C. Campbell and M. Bisson. Anal. Chem. $\underline{51}$: 844-851 (1979).

13. Slavek, J. and W.F. Pickering. Water, Air & Soil Poll. $\underline{28}$: 151-162 (1986).

Section 2

Biological Implications of Metal Speciation

TESTING THE BIOAVAILABILITY OF METALS IN NATURAL WATERS

Peter Pärt

Department of Zoophysiology, Uppsala university

Box 560, S-751 22 Uppsala, Sweden

Introduction

Natural waters show a great diversity in their physical and chemical properties. The various mechanisms by which water quality controls the availability of metals to aquatic organisms is therefore of parmount importance when we want to predict metal impacts in the aquatic environment. The basic question to answer is; to what extent does a measured water concentration of a metal relate to the amount taken up by the organisms? The present article is an attempt to review some of the methods that have or could be used in measuring the biological availability of metals. Two approaches can be distinguished. The first is methods that aim to identify bioavailable forms or fractions of metals as a guide to the development of methods for chemical analysis of these enteties. The second is methods to be used in the evaluation of the actual bioavailability in the prevailing conditions in a natural water.

The bioavailability of a chemical is a relative concept. The availability varies depending on trophic level of the organism, developmental stage, feeding strategy etc. Moreover it depends on the form of occurence, size, lipophilicity, polarity and charge of the molecules under consideration. However, ultimately the bioavailability of a xenobiotic is determined by the physical and chemical properties of the barries which the organisms expose to the environment. These barriers exhibit various degrees of complexity in structure and function. They include compartmental epithelial cell systems as well as simple structures like the unit membrane that determines the morphological boundries of the cell. The mechanisms by which xenobiotics are bound and transported through cellular membranes is poorly understood, although passage of membranes and binding to ligands are two phenomena fundamental in biological transfer and accumulation.

So, the key physiological process when discussing metal bioavailability is the mechanism of uptake. It is at this level that we have the sieve which determines if a particular metal form or fraction is available or not. Uptake is defined in the OECD Guidelines for toxicity testing as "the process of sorbing a test compound into and/or onto the organism". The uptake rate is therefore the natural measure of bioavailability. To be considderd bioavailable, a metal species or form must be taken up with a rate that later on will result in a net accumulation of the element, i e the uptake have to exceed exkretion and dilution within the body due to growth.

Until now, direct measurements of uptake rates is not commonly applied in bioavailability studies with metals. The reason is obvious - lack of reliable methods. Instead, bioavailability is evaluated from toxicity or bioconcentration data. Although these approaches have contributed with valuable information, it must be kept in mind that response parameters are indirect measures of bioavailability. Besides the uptake rate, they depend on factors such as binding, excretion, metabolism and toxic mechanism. Therefore, they do not permit straight conclusions about availability except in well defined and controlled experimental situations. So, it is to be preferred in the future that bioavailability studies will include direct measurements of uptake rates across the borders which aquatic organisms expose to the environment.

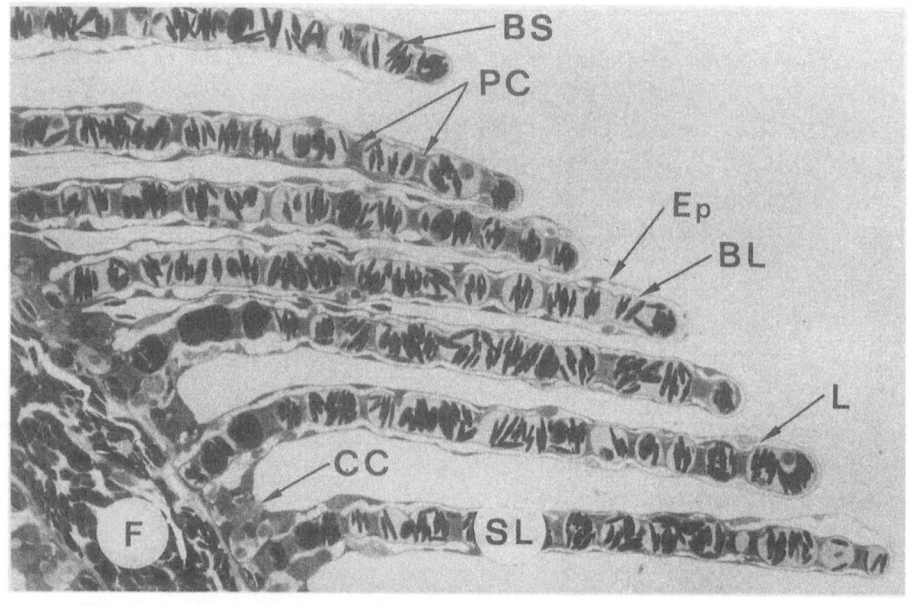

Figure 1. Longitudinal section of a gill filament from a freshwater adapted rainbow trout. 1 um section stained in toluidine blue.
F = gill filament
SL= secondary lamellae
BS= blood space containing erythrocytes
PC= pillar cells with flanges lining the blood spaces
EP= epithelium of flat epithelial cells joined by tight junctions
CC= Chloride cells at lamellar bases
BL= basal lamina
L = lymphatic space

Physiology of the barriers

Aquatic organism, except unicellular alage, are covered with epithelia towards the external environment. The epithelium consists of epithelial cells in single or multiple layers, with the cells joined by regions of apparent membrane fusion - the tight junctions. Substances penetrate the epithelium either by a trans-cellular route through the cells, or by a para-cellular route via the junctions. The thickness and properties of the epithelia varies between different parts of the organism and further on between organisms. Generally, the thinnest epithelia are found in the respiratory organs (gills), which are specialized and optimized for gas exchange with the environment (large surface area, short diffusion distances between blood and water). A representative example is the fish gill (fig. 1).

In general the permeability properties of epithelia are similar to those of cell membranes. The membrane consists of a bi-molecular layer of phospho-lipid molecules perpendiculary orientated to the plane of the membrane with their polar parts towards the surface and the non-polar parts inwards. Proteins tranversing the thickness of the membrane are embedded in the lipoid layer. These proteins act as comunicators between the interier and exterior of the cell by participating in the translocation of ions and nutrients across the membrane.

The mechanisms for heavy metal penetration through the epithelial cell membranes is basically unknown. However, the following have been suggested (1):
1. Diffusion in non-polar complexes through the lipoid regions of the membrane.
2. Penetration in ionic form by binding to carrier proteins in the membrane, "facilitated diffusion".
3. Endocytosis of particulate fractions. A vacuole is formed from the cell membrane around the particle. The whole package is engulfed by the cell and digested intracellularly (particulary important suspension feeders).
4. Metals complexed to nutrients (amino acids, carboxylic acids) by nutrient specific carriers.

Water quality may affect the penetration rate of metals, e g bioavailability, by two fundamental processes. Firstly, by modifying the chemical speciation of the metal in the water, hereby making it either more or less available for the uptake mechanism. For example, soluble ligands that binds the metal ion in the water will prohibit binding to any type of "carrier" in the membrane thus decreasing the uptake rate. However, depending on the nature of the ligand and the chemical properties of the metal complex formed - polar or non-polar - complexation could also increase the availability. The uptake rate of cadmium through fish gills was shown to increase in the presence of xanthates, probably because of the formation of non-polar, lipophilic metal-xanthate complexes (2). Secondly, water quality may change the availability by a biological mechanism by changing the properties of the epithelia. An example is the commonly observed antagonistic effect of Ca^{2+} on metal accumulation and toxicity which most likly is the result of Ca^{2+} decreasing the epithelial permeability of the metals (3).

Methods to identify bioavailable forms

The following nomenclature for the chemical speciation of metals is used (4). Metal species refer to those enteties (ions, molecules, complexes) which can be described in terms of well defined stoichiometry. Metal forms include both species and less well defined enteties (e g metal associated with un-characterized organic material). Finally, metal fraction refers to groups of forms resolved by - and thus operationally defined by - particular analytical techniques.

Unicellular algae grown in culture have widly been used as monitors of chemical speciation in metals (5,6,7,8,9,10,11). This approach has proven to be successfull although response parameters are used. Examples of proper bioavailability studies with algae are those of Guy and Ross Kean (12) and Florence et al.. (13). Both have used algal growth rate as a response parameter to investigate the availability of copper in the presence of various natural and artificial organic ligands. Growth was evaluated from the number of cells or, better, chorophyll or ATP content. The assays include careful chemical speciation of the metal by filtration, anodic-stripping voltametry, ion selective electrodes and binding to ion exchange resins (Chelex-100). These measurements were combined with theoretical calculations of metal speciation from existing equilibria, and the toxicity of various metal species, forms and fractions could be compared.

The results points on the free metal ion (Cu^{2+}) as the principally available metal species while complexed metal generally was less available. However, they also found several exceptions where complexed metal proved to be toxic. Here a drawback of using response parameters is illustrated. Although complexation increased metal toxicity, it does not necessarily mean an increased availability. The possibility of the metal complex being more toxic as such, than the free metal ion, has first to be investigated before conclusion about bioavailability can be made.

A way to overcome this uncertainity is direct measurements of the uptake rates in the algae. It is preferably done by using radioactive isotops, because analysis of radioactivity is much simpler than conventional analysis of stable elements. The algal cells are exposed to the radioactive metal in the incubation medium. Cells are sampled at regular time intervals and the cell-associated radioactivity is measured. The exposure should ideally continue until an apparent "steady-state" is reached. The initial uptake rate is calculated from the accumulation curve by non-linear regression (14,15). The obvious advantage of this approach is direct measurements of the bioavailability of various metal species or forms. Moreover, the total metal concentration can be kept low - often at more environmentally realistic concentrations than in toxicity tests - by the use of radionuclides with high specific activity. Possible toxic effects on the mechanisms which determines the bioavailability in the cells is herby also eliminated. Unfortunately, this approach has not sofar been appreciated in bioavailability studies with algae, except in the study of Macka et al. (16) where the uptake of mercury and cadmium in the microalgae Chlamydomonas reinhardi WT[+] was investigated.

Crustaceans grown in laboratory culture under controled conditions for many generations have also been used in bioavailability studies of metals (17,18,19,20, 21). Here again effect parametars such as survivial, reproductive success or tissue content has been used in the evaluation. Particulary in the studies of Sunda et al. (22), Poldoski (23) and Borgman and Ralph (24) a thourough chemical speciation of the metals (cadmium and copper) is combined with their toxicity or tissue concentrations. Metal speciation is based on measurements with ion selective electrodes and theoretical calculations. As found with the alagal tests, the free metal ions proved generally to be the most available species. Inorganic and organic complexation decreased metal availability. However, similarily to the algae, increased availability was observed with certain organic ligands. Low molecular weight natural organic material (18), sodiumdiethyldithiocarbamate (23) and ethylxanthogenate (25) were found to increase toxicity or tissue accumulation. The inclusion of radionuclides in these assays for calculation of initial uptake rates and to avoid toxic metal concentrations is to be recomended as discussed previosly.

Direct measurements of the uptake rates across the epithelia of aquatic organisms is a desirable future development of bioavailability studies. Unfortunately measurements in intact animals appears to be unrealistic. However, in vitro preparations of isolated organs or epithelia may be useful in this respect. One such approach is the isolated perfused fish gill preparation (Figure 2). Basically, the gills are dissected from the fish body and the afferent and efferent blood vessels are cannulated. The efferent cannula is connected to a perfusion pump delivering the perfusion medium, usually physiological saline containing albumin or dextran as plasma protein substitute. After passage of the gills, the perfusion medium is collected from the efferent branchial vessels. The gills are submersed in water and effectively irrigated during perfusion. This type of gill preparation allows direct measurements of uptake rates of substances from the water to the perfusion medium, or vice vers from medium to the water (excretion). Until now it has basically been used in studies of gill physiology (26), but it appears also to have an applicability in bioavailability studies.

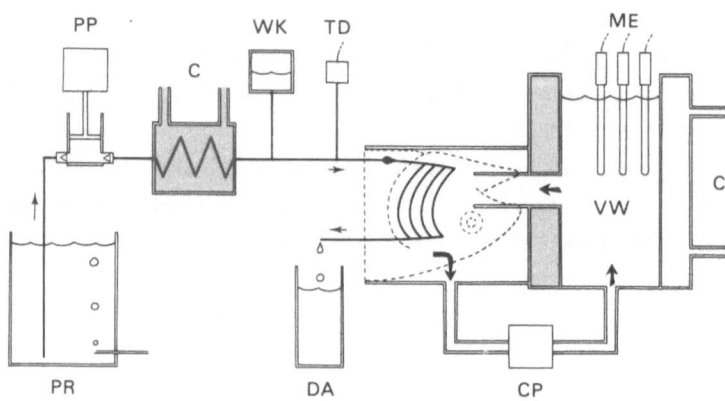

Figure 2. Experimental set-up for perfusion of fish gills used in cadmium uptake experiments.
PP= perfusion pump, replaces the heart
CP= circulatory pump ventilating the gills
C = circulating refrigerant from thermostat unit
WK= pulse damper
TD= pressure transducer for registration of perfusion pressure
ME= ion selective electrodes for measurement of Na^+, Cd^{2+}, pH
VW= ventilatory water
PR= perfusion medium
DA= efferent perfusate collected from the dorsal aorta after passage of the gills

The uptake of cadmium in the perfused gill preparation from freshwater adapted rainbow trout (<u>Salmo gairdneri</u>) was investigated in different water hardnesses, salinities and pH´s (3), in the presence of organic complexants (EDTA and citrate)(27) and diethyldithiocarbamate, ethyl- and isopropylxanthate (2) and in the presence of detergents (NP10-EO and LAS)(28). It was found that Cd uptake basically was a function of free Cd^{2+} in the water, that Ca^{2+} decreased uptake, that Cd-citrate complexes to some extent were available and that the anionic tensid LAS facilitated Cd uptake. Moreover, the dithiocarbamate and the xanthates increased Cd uptake The method has also been used in studies of the availability of hexavalent chromium at two pH levels (29). One conclusions from these perfusion studies is that the chemical properties of the metal complexes are of importance with respect to their bioavailability. Three categories could be distinguished. The first comprises metal complexes un-available to aquatic organisms type EDTA (ethylenediaminotetraacetic acid), NTA (nitrilotriacetic acid) and DTPA (diethylenetriaminopentaacetic acid).

The second includes complexes that are available to some extent, but the complexed metal is still less available than the free metal ion. Within this groupe low molecular weight fulvic acids, amino acids and carboxylic acids (citrate) are found. The third containes non-polar metal complexes with a higher availability than the free metal ion. Examples in this groupe are diethyldithiocarbamates and xanthates.

An alternative method for measuring uptake rates across epithelia is the "Ussing chamber technique" (30). The prerequisite is that the epithelial tissue can be mounted flat. Is is clamped between two chambers thus separating the solutions in two environmental compartments. The metal, preferably as a radioactive nuclide, is added to one compartment and the appearence in the opposite compartment is followed by regular sampling. This technique has mostly been used in studies of ion permeability of various epithelia like the frog skin, the toad urinary bladder, the gall bladder, the intestine of mammals and fish and fish skin. It has not until now been used in bioavailability studies, neither of organic xenobiotics or of metals, but appears to have a potential in this kind of studies as well.

Methods for evaluating bioavailability under natural conditions

One approach to estimate the bioavailable concentration of metals under field conditions is to use indicator organisms. The most frequently used are mussels (Mytilus edulis) and Fucus sea-weeds. The critical assumption in using idicator organisms is that they accumulate the metal in direct proportion to the environmental concentration. To what extent this criteria is fullfilled is many times still a question of debate. Anyhow, Bryan (31) concludes that analysis of Fucus vesiculosus probably gives a good indication on the avarage bioavailability of silver, cadmium, copper, lead and zinc in waters modified by factors including inorganicand organic complexation, presence of particulate forms and competition from other metals.

A method to estimate the bioavailable concentration of lead in sediments is presented by Luoma and Bryan (32). They found a good correlation between the Pb concentration in the deposit feeding bivalve Scrobicularia plana and the lead/iron ratio in 1 M hydrochloric acid extracts of sediment (Figure 3). Similarily, Langston (33) tried successfully the same concept on the arsenic content in the same animal but extended the studies to also include the polychaet Nereis and Fucus. The conclusion is that Pb/Fe and As/Fe ratios in acid extracts of sediments is a good predictor of Pb and As bioavailability to the deposit feeder but not to Nereis and Fucus.

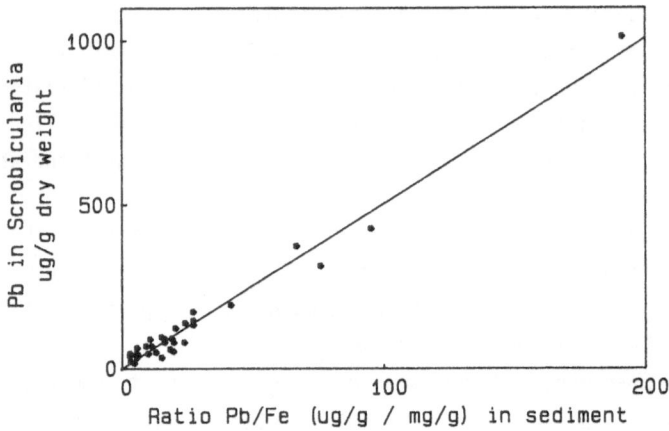

Figure 3. Correlation between concentrations of lead in soft tissues of Scrobicula-
          ria plana and ratio Pb/Fe extracted with 1 N hydrochloric acid from sedi-
          ment (after ref. 32)

A promising approach to measure the bioavailablity of metals to fish under
natural conditions is presented by Bendell-Young et al. (34). They used the Cd content in the liver and the growth rate of the fish to compare Cd bioavailability in 6
acidified lakes. The amount of Cd in the liver was assumed to be a function of the
bioavailability and the weight of the fish:

$$Cd_{liver} = f \text{ (bioavailability, weight)}$$

They measured the liver Cd content, the age from examining the scales and the
weight. From these measurements they could estimate the change in liver Cd content
over time (dCd/dt)(mg Cd/year) and the absolute change in weight over time
(dW/dt)(gram/year). Assuming that the bioavailability not is changing over the year,
then the change in liver Cd content over time to growth rate is a function (Z) of
bioavailability.

$$dCd/dt \ / \ dW/dt = Z \text{ (bioavailability, weight)}$$

A plot of dCd/dt / dW/dt versus W for the various fish populations will yield
straight lines, where differences in the slopes reflects differences in bioavailability (figure 4). Analysis of the slopes in figure 4 shows that the bioavailability
of Cd was essentially the same in four of the lakes. In two lakes, marked with 1
and 2 in fig. 4, the Cd availability was comparably reduced. Lake nr. 1 had the highest pH (pH=6.4) of the lakes investigated, while lake nr. 2 appeared to be unique
among the lakes in that it was a small, shallow and highly organic, acidic lake. The
authors assume Cd to be less available in this lake because the metal is bound to
organic material.

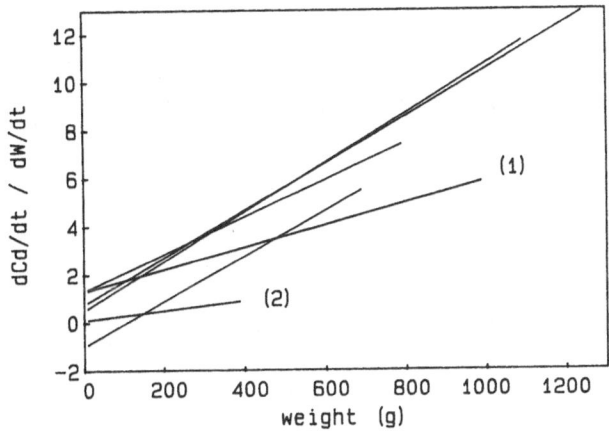

Figure 4. Relative bioavailability factors. Lake nr. 1 has the highest pH of the lakes (pH=6.4). Nr. 2 is acidic buy highly organic (after ref. 34).

As a conclusion, simple methods for measuring bioavailability which are universal to all organisms will certainly not be found. Rather, the solution to the bioavailability problem lies in a basic understanding of the physiological, biochemical, geochemical and ecological controls on the process. Such deepend understanding will be the key to the future development of both chemical and biological methods to be used in metal speciation and bioavailability studies. The natural measure of bioavailability is the uptake rate. By the use of radionuclides, direct measurements of uptake rates could be made in those biotests with alga and crustaceans today used bioavailability studies. The perfused gill and similar in vitro preparations, which allow direct measuremets of uptake rates across the epithelia exposed, may provide further information on the bioavailability of particular metal species, forms and fractions. For the field situation several methods, or approaches are available. Among those, the recently presented one (Bendell-Young et al. 1986), where liver metal content is related to growth rate appear promising and deserve attention in the future.

Acknowledgements

Financial support has been obtained from the National Swedish Environment Protection Board within the project areas Fish/Metals and ESTHER.

References

1.   Luoma, S. The Sci. Total Environm. 28: 1-22 (1983).

2.   Block, M. and P. Pärt. Aquat. Toxicol. 8: 295-302 (1986).

3.   Pärt, P., O. Svanberg and A. Kiessling. Water Res. 19: 427-434 (1985).

4.   Burton, J.D. Phil. Trans. R. Soc. London B. 286: 443-456 (1979).

5.   Davey, E.W., M.J. Morgan and S.J. Erickson. Limnol. Oceanogr. 18: 993-997 (1973).

6.   Gnassia-Barelli, M., M. Romeo, F. Laumond and D. Pesano. Mar. Biol. 47: 15-19 (1978).

7.   Gächter, R., J.S. Davies and A. Mares. Environm. Sci. Technol. 12: 1416-1421 (1978).

8.   Sunda, W.G. and J.A.M. Lewis. Limnol. Oceanogr. 23: 870-876 (1978).

9.   Allen, H.E., R.H. Hall and T.D. Brisbin. Environm. Sci. Technol. 14: 441-443 (1980).

10.   Rai, L.C., J.P. Gaur and H.D. Kumar. Environm. Res. 25: 250-259 (1981).

11.   Petersen R. Environm. Sci Technol. 16: 443-447 (1982).

12.   Guy, R.D. and A. Ross Kean. Water Res. 14: 891-899 (1980).

13.   Florence, T.M., B.G. Lumsden and J.J. Fardy. Anal. Chim. Acta 151: 281-295 (1983)

14.   Borle, A.B. Cell Calcium 2: 187-196 (1981).

15.   Brahm, J. J. gen. Physiol. 82: 1-23 (1983).

16.   Macka, W., H. Wihlidal, G. Stehlik, J. Washuttl and E. Bancher. Chemosphere 9: 409-413 (1979).

17.   Andrew, R.W., K.E. Biesinger and G.E. Glass. Water Res. 11: 309-315 (1977).

18.   Giesy, J.P., G.J. Leversee and D.R Williams. Water Res. 11: 1013-1020 (1977).

19.   Blust, R., E. Verheyen, C. Doumen and W. Decleir. Aquat. Toxicol. 8: 211-221 (1986).

20.   Winner, R.W. Aquat. Toxicol. 8: 281-293 (1986).

21.   Winner, R.W. and J.D. Gauss. Aquat. Toxicol. 8: 149-161 (1986).

22.   Sunda, W.G., D.W. Engel and R.M. Thuotte. Environm. Sci. Technol. 12: 409-413 (1978).

23.   Poldoski, J.E. Environm. Sci. Technol. 13: 701-706 (1979).

24.   Borgmann, U. and K.M. Ralph. Water Res. 11: 1697-1703 (1983).

25.   Ahsanullah, M. and T.M. Florence. Mar. Biol. 84: 41-45 (1984).

26. Perry, S.F., P.S. Davie, C. Daxboeck, A.G. Ellis and D.G. Smith. In: W.S. Hoar and D.J. Randall, Eds., Fish Physiology, Vol XB, pp. 326-388. (Academic Press, 1984).

27. Pärt, P. and G. Wikmark. Aquat. Toxicol. 5: 277-289 (1984).

28. Pärt, P., O. Svanberg and E. Bergström. Ecotoxicol. Environm. Safety 9: 135-144 (1985).

29. van der Putte, I. and P. Pärt. Aquat. Toxicol. 2: 31-45 (1982).

30. Ussing, H.H. and K. Zerahn. Acta Physiol. Scand. 23: 110-127 (1951).

31. Bryan, G.W. The Sci. total Environm. 28: 91-104 (1983).

32. Luoma, S. and G.W. Bryan. J. mar. biol. Assoc. U.K. 58: 793-802 (1978).

33. Langston, W.J. J. mar. biol. Assoc. U.K. 60: 869-881 (1980).

34. Bendell-Young, L.I., H.H. Harvey and J.F. Young. Can. J. Fish. Aquat. Sci. 43: 806-811 (1986).

CASE STUDIES ON METAL DISTRIBUTION AND UPTAKE IN BIOTA

Olle Grahn and Lars Håkanson
Swedish Environmental Research Group (SERG)
Fryksta 665 00 Kil Sweden
Dept. of Hydrology, University of Uppsala
V Ågatan 24 752 20 Uppsala

ABSTRACT

The aim of this paper is to sum up some results on heavy metals from
Swedish studies on the linkage between metal contamination, environmental
sensitivity, biological uptake and ecological effects.

Man's activity has caused a drastic increase of metal fluxes in the
biosphere by emissions from industries and by burning of fossil fuels.
The increased acidification of soils and waters has also a strong bearing
on the distribution and fate of metals and the mobilization of metals
from soils.

The bioavailable part of the metals present in natural waters varies
within a wide range due to several environmental factors. Metals are
taken up differently by different test organisms.

In future evaluations of different kinds of metal pollution in different
types of aquatic environment, it is necessary to make metal speciations
as a step to determine which fractions are bioavailable.

INTRODUCTION AND AIM

Numerous papers and reports have been written on the transport, turnover
and ecotoxicological aspects of metals in aquatic systems. Excellent
general surveys have been given by, e.g., Förstner and Wittman (1979) and
Salomons and Förstner (1984).

The aim of this paper is to sum up some results from Swedish studies on
the linkage between metal contamination, environmental sensitivity,
biological uptake, ecological effects and assessment of heavy metals

in limnic environments, and to concentrate on some case studies of
principle interest. For a more thorough discussion on the causal
relationships between dose, environmental sensitivity and ecological
effects of metals, see Håkanson (1984a, 1984b). Thus, here we do not
discuss methods, reliability of data, etc., but will focus on results of
general, principle interest. Most examples will be taken from
investigations in areas affected by discharges from mines and acid
deposition. We will not try to give a general literature review. These
presuppositions imply that the reference list gives a distorted and poor
picture of the most important publications on metals in aquatic systems.

Some metals like Fe, Mn, Cu, Zn, Co and Mo, are essential for life in the
proper concentration range. This range can be narrow. Other metals, like
Hg, Cd and Pb are not essential but are toxic to smaller or larger
extents depending on concentration and environmental conditions. In this
paper we will address some conditions of importance for aquatic
ecosystems.

Man's activity has caused a drastic increase of metal fluxes in the
biosphere by emissions from different types of industry, e.g., mines,
smelters and iron- and steelworks. Emissions of acidifying compounds and
metals to air, water and soils by burning of fossil fuels have also
increased significantly during recent decades. This has a strong bearing
on the distribution and fate of metals and also causes mobilization of
metals in the soils.

OLD SINS AND NEW

In the area of Garpenberg in central Sweden (Fig. 1) mining activities
have been going on since the Middle Ages. Today area is one of the most
metal-contaminated areas in Sweden. Roads and entire communities are
partly built on wastes from the mining. Fig. 2 gives a location map
illustrating the sites of the present mines, locations of major older
deposits and the lake (Gruvsjön or Mine lake in translation) which
recieves a heavy load of many metals.

The diagram in Fig. 2 differentiates between the metal fluxes of Cu, Pb,
Zn and Cd from the following sources:

- diffuse sources around the lake
- old deposits

- present mines

It should be noted that for Cu, Zn and Cd only a small percentage, of the
total metal flux to Lake Gruvsjön today emanates from the present mining
activity. For Pb about 40% can be linked to present industries
(Lindeström and Qvarfort 1985)

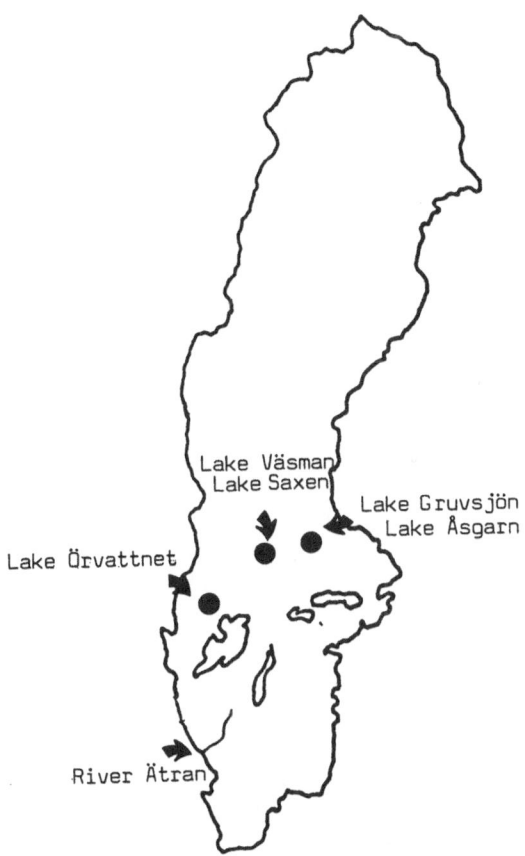

Figure 1: Location map of discussed lakes and rivers.

It is evident that significant amounts of metals are transported to Lake
Gruvsjön. This has caused high metal concentrations in water, sediments
and all types of investigated organisms and a reduced bottom fauna and
fish population in the lake, which is oligotrophic. In Lake Åsgarn,
however, which is located about 5 km downstream from Lake Gruvsjön and
which has a much higher level of bioproduction due to nutrient load the

concentrations of metals in biota are significantly lower and no apparent
ecological effects have been documented - in spite of the fact that the
metal load also to this lake is very high (Grahn and Sangfors 1984).

Acidification is a well documented menace in many parts of Scandinavia
and North America. The spread and ecological effect of many metals are
highly related to the pH of the water. It has been demonstrated that the
leakage of metals from soils increases significantly with lower pH. A
clear example of this is given for Cd in Fig. 3. Note the marked peak
below pH 5.

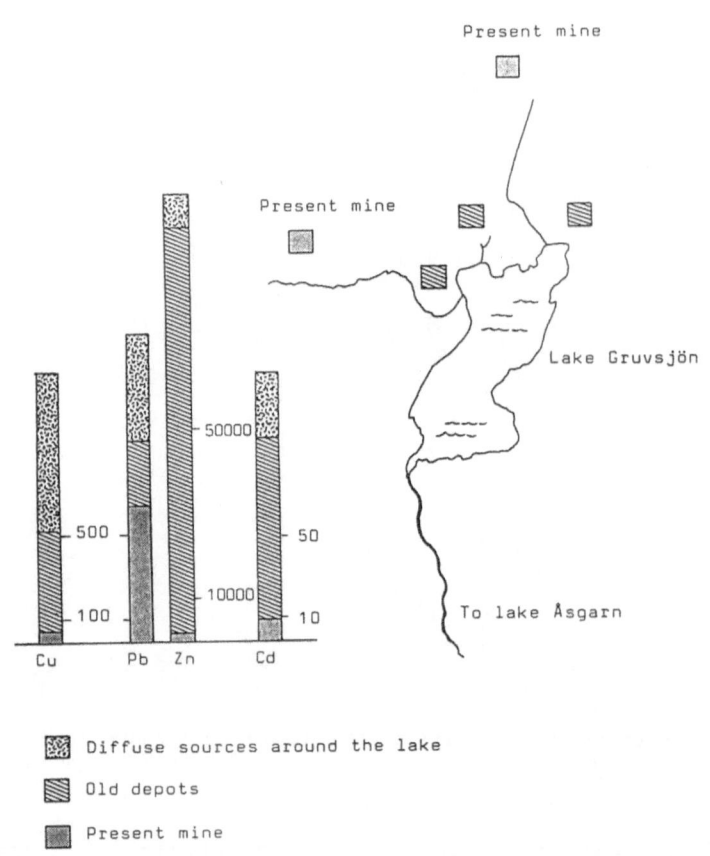

Figure 2: Location map of the area around Lake Gruvsjön. Estimated fluxes
of metals (in kg) from May 1979 to April 1980.

An investigation in different watersheds in Sweden where mean pH in the run-off water varied between 4.1 -5.0 also clearly demonstrates that the transport of metals is much higher in areas with low pH (Fig. 4). The mobilization of, e.g., Zn, Cd, Al and Co is 5 - 10 times higher in the area with pH 4.1 in the run-off water than in the area with pH 5.0 (Grahn and Rosén 1984).

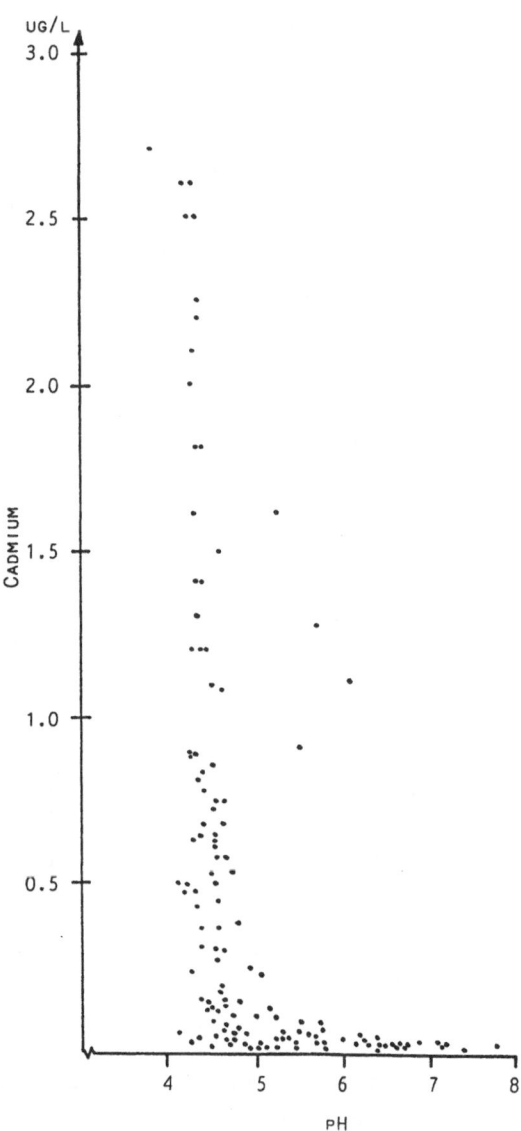

Figure 3: The relationship between pH and Cd concentrations in lake and soil water.

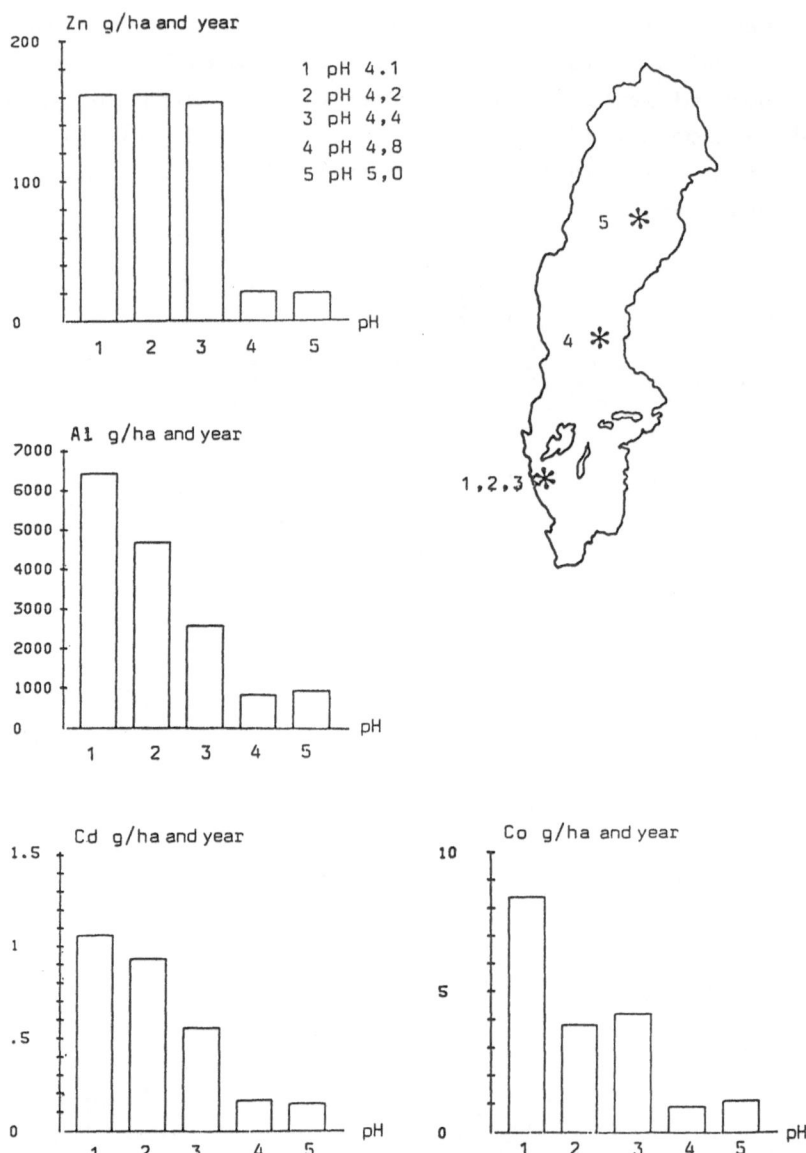

Figure 4: Mobilization of metals in areas with different pH in the drainage water (g/ha and year).

To conclude: Old mine deposits can leak significant amounts of metals, mainly due to sulfide oxidation. The potential ecological damage would depend on the character of the receiving water, e.g., the trophic status as well as the metal dose. Metal mobilisation will also take place in acidified areas and the mobilization is negatively correlated to the pH.

Lake	Area (ha)	Volume (Mm^3)	Mean depth (m)	Mean pH	Trophical level
Gruvsjön	141	4.2	3	7.0	Oligotrophic
Åsgarn	183	3.7	2	6.7	Eutrophic
Saxen	82	2.5	3	6.6	Mesotrophic
Väsman	3860	410	10.6	6.8	Mesotrophic
Örvattnet	72	6.0	8.4	4.9	Oligotrophic

Table 1:   Background data on lakes discussed.

Lake	Hg		Cd		Pb		Cu		Zn	
	wat	sed	wat	sed	wat	sed	wat	sed	wat	sed
Gruvsjön	–	1.5	5.6	25	12	9100	20	4800	3500	11200
Åsgarn	–	1.6	1.6	40	4	900	6	860	1200	10400
Saxen	–	0.95	–	105	20	16100	150	4030	2400	35430
Väsman	–	2.35	–	5.1	–	790	–	100	–	1210
Örvattnet	–	–	0.3	3	2	210	2	28	10	300

Table 2:   Metals in water and surficial sediment of lakes discussed in ug/l and ug/g ds.

THE NEED FOR DIFFERENTIATION

Heavy metals are a most inhomogeneous group and the sensitivity of receiving waters to metal contamination may vary within a wide range.

Lake Saxen receives waste water from a mine, Saxberget. Mining has been going on in this area since 1880 the present industry starting in 1936. Very high concentrations of most metals are found in water and sediments (Table 2).

Lake Väsman, which is linked to Lake Saxen via a short river, receives smaller amounts of most of the metals (Lindeström 1984).

Comparing the situation in these two lakes, how are the metals picked up in biota?

We will illustrate some central concepts with data on Zn and Cd in
plankton, perch muscle and perch liver (Fig. 5). From these diagrams it
is clear that:

- Zn is found in much higher concentrations than Cd in the test organisms

- Very little Zn and Cd are found in fish muscle.

- The Zn concentrations are much lower in plankton in Lake Väsman than in
  Lake Saxen - but the Zn concentration in perch liver and muscle is
  about the same in the two
  lakes.

- The Cd concentration is about the same in the plankton in the two lakes
  - but the Cd concentration is much lower in perch liver in Lake Väsman.

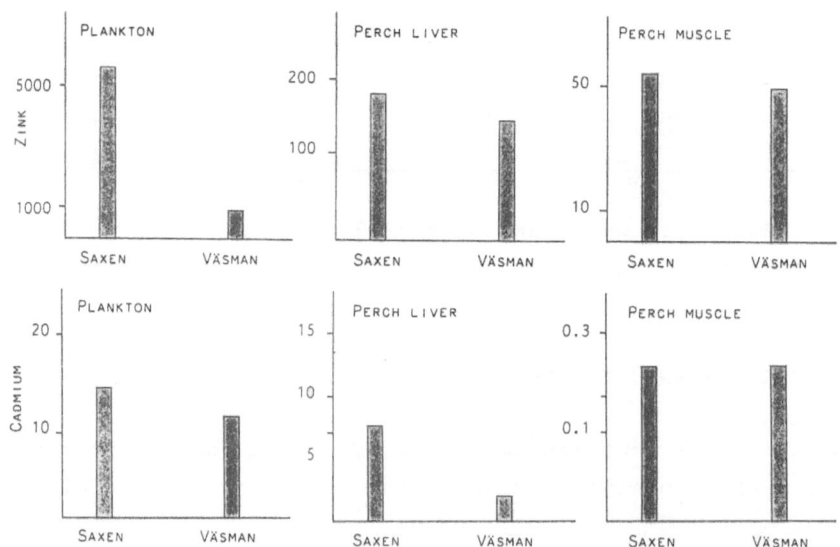

Figure 5: Zinc and cadmium in plankton, pike liver and perch muscle in
           Lake Saxen and Lake Väsman.

Confusion would prevail if we continue to determine total concentrations
of metals. The crucial question is: How much of a given metal dose is
biologically available and can be picked up in the food chains?

This is a most complex issue. It is also important to note that different metals are taken up differently by different test organisms. Fig. 6 illustrates, for Zn and Cd, that:

- different test organisms react differently to one and the same metal - the range causing long-term and short-term effects (for Zn and Cd) is often more than 3 orders of magnitude in concentration;

- Zn and Cd (just as most metals) do not biomagnify (i.e. the concentration is not higher in the predator than in its prey). Hg is an exception from this general rule for "heavy metals" since Hg appears in the form of methyl mercury which is lipophilic. We will return to Hg later on.

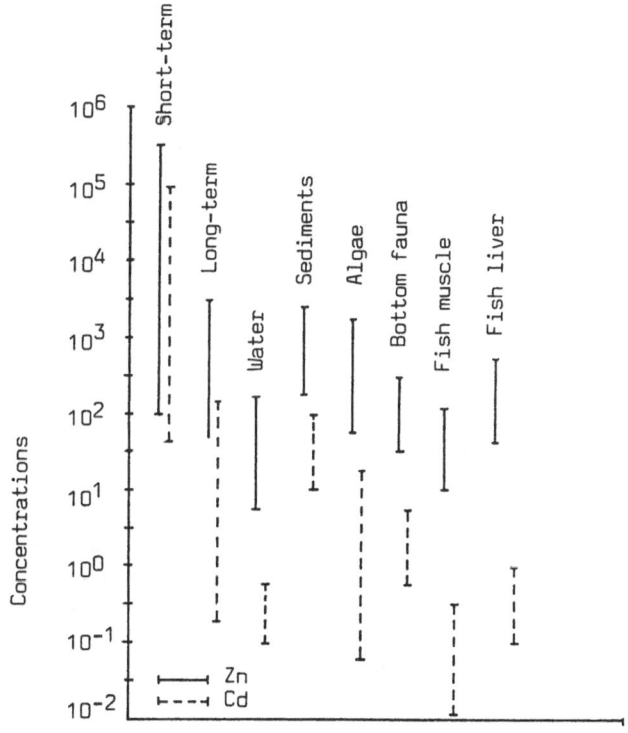

Figure 6: Concentrations of Zn and Cd yielding short-term and long-term effects, and concentrations in water, lake sediments, attached algae, bottom fauna, fish muscle and fish liver (from Håkanson and Jansson, 1983).

One crude way to try to distinguish the biologically available dose from the total dose is to make a fractionation. Many fractionation (=speciation) schemes exist (see, e.g., Chester and Hughes, 1967; Tessier et al. 1979; Salomons and Förstner, 1980).

METAL SPECIATION

Most metals can appear in different chemical forms with different characteristics and different affinities to various types of "carrier particles" (Fig. 7) existing in natural water systems (humic substances, Fe/Mn-oxides, carbonates, etc.). These "carrier particles" may camouflage the toxic properties and govern the spread in nature in a way that is often very difficult to foresee from laboratory tests.

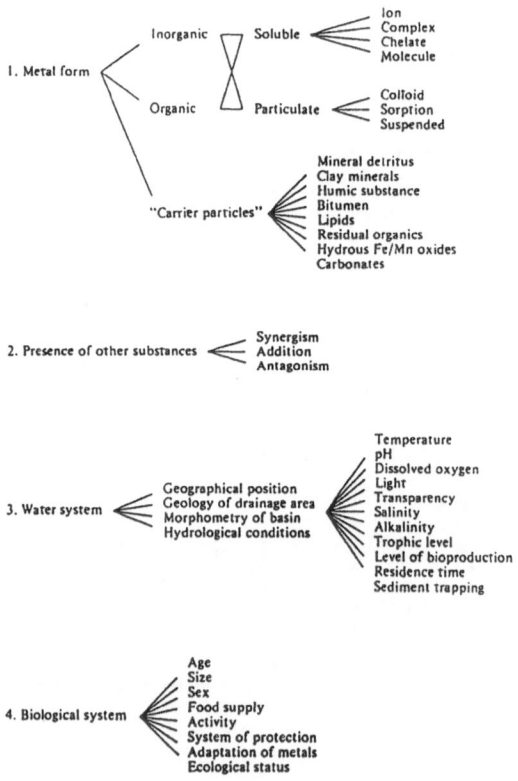

Figure 7: Factors affecting the toxicity of metals in aquatic systems. From Beijer et al. (1977) and Håkanson and Jansson (1983).

Fig. 8 exemplifies the distribution of various metal fractions in
suspended matter and surficial sediments in the river system of the River
Ätran (see Fig. 1 for geographical orientation). The following fractions
have been determined (Landner & Grahn, 1975).

- Metals (Cd, Zn, Pb, Cu, Cr, Ni, Fe and Mn) adsorbed on particles. This
  may be called the easily exchangeable fraction.

- Metals co-precipitated as hydroxides or carbonates on particles. Metals
  in this fraction are less available for biological uptake than metals
  in the first group.

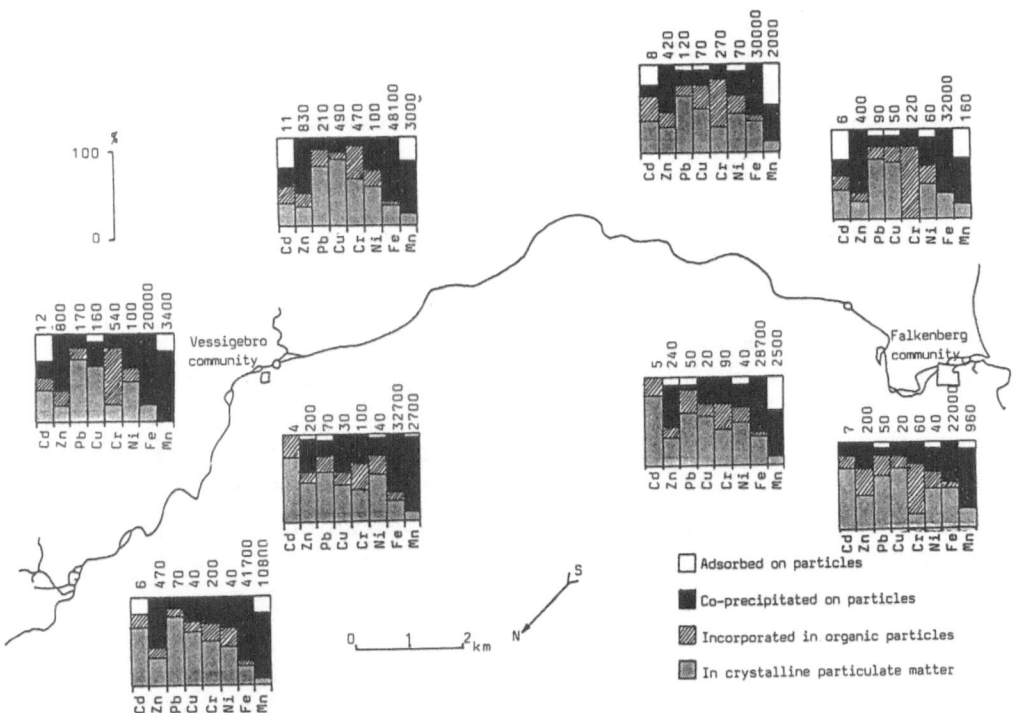

Figure 8: Various fractions of selected metals in sediments and
suspended matter in the River Ätran, southern Sweden.

- Metals incorporated in organic particles. These metals may be picked up
  by organisms feeding on organic matter or when the organic aggregates,
  the "carrier particles", are mineralized.

- Metal in crystalline particulate matter. Metals in this fraction are the least biologically available and may be called the inert fraction.

From the figure we may note that:

- 50 to 80% of the Cd, Pb and Cu appear in the inert fraction in the surficial sediment and 50 to 80% of Pb and Cu in the suspended matter.

- The easily exchangeable fraction is small for most toxic metals, except for Cd where 25% - 50% belongs to the fraction.

The results from this water system concerning the affiliation of the various metals to the various fractions cannot be applied to other water systems with different characteristics in terms of, e.g., pH, alkalinity, salinity or humic content, since these environmental factors strongly affect the distribution of the metals to the various fractions.

ENVIRONMENTAL SENSITIVITY

The chemical, limnological, sedimentological and hydrodynamical characters of the receiving waters influence the spread and fate of metals (Fig. 7). Subsequently, we will give one typical example of this.

The objective of this particular investigation (Grahn and Sangfors, 1984) was to study the seasonal variability of selected "heavy metals" in perch, zooplankton and water in four lakes which varied in metal contamination, acidity and trophic status (Tables 1 and 2). Results will be discussed from two of these lakes:

- Lake Åsgarn; high metal pollution, high pH and high bioproductivity;

- Lake Örvattnet; low metal pollution, acidified with pH ´5, low bioproductivity.

Fig. 9 shows Cd and Zn concentrations in water and perch liver from May 1983 in these two lakes. Zn concentrations are generally much higher than Cd concentrations. The most remarkable is that the Cd concentration in the water is about 6 times lower in the "unpolluted" lake, but the Cd concentration in fish liver is about 3 times higher.

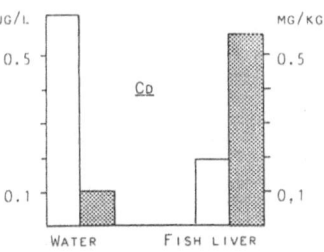

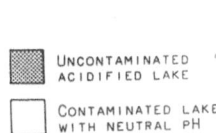

Uncontaminated
Acidified lake

Contaminated lake
with neutral pH

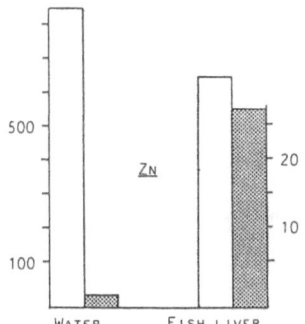

Figure 9: Zinc and cadmium concentrations in water and
perch liver from Lake
Åsgarn and Lake Örvatt-
net, May 1983.

How can this occur? Once again we must seek the answer in the available
form of the metals and the environmental factors that govern this form.
In the acidified and oligotrophic lake, more of the cadmium is in the
easily available form, and the biological dilution is lower due to the
lower amount of suspended matter (plankton, humus substances, etc.).
Thus, total analysis of metals may be a crude approach in assessments of
potential biological risks.

MERCURY

About 40 000 papers have been published on mercury in the environment
(Andersson et al., 1987). Mercury may be regarded as a special case among
the "heavy metals" due to its toxicity, ability by bioaccumulate and
because of its threat to man (consuming fish with high Hg content). Lakes
can be declared "black-listed" if the Hg content in fish (1 kg pike)
exceeds 1.0 (mg Hg/kg) in Sweden and 0.5 mg Hg/kg in Canada. Most mercury
in fish appears in the form of methyl mercury, a most toxic variety.

The mercury content in fish (1 kg pike) depends on many factors, e.g.,
(Håkanson, 1980):

- the dose of Hg to the lake;

- the acidity of the water; the lower the pH the higher the Hg content in fish;

- the trophic level of the lake; the higher the lake bioproductivity, the lower the Hg content in fish.

The relationship between these factors may be given by the formula:

$$F(Hg) = (4.8*\log(1 + Hg50/200))/((pH - 2)*\log (BPI))$$

where

$F(Hg)$ =   the content methyl mercury in muscle in 1 kg pike in mg/kg. This may be regarded as a effect term or a biological response parameter.

Hg_{50} =   the areal median Hg content of surficial sediments (0 - 1 cm, in ng/g ds). This is one example of a dose parameter.

pH   =   the mean pH of the water system. This is an example of a lake sensitivity parameter.

BPI   =   the bioproduction index (Håkanson, 1984b). This is a second lake sensitivity parameter.

A comparison between model data from the given formula and a set of empirical data has revealed very good correlation; the correlation coefficient, r, is 0.86 (Table 3).

If two sets of empirical fish data from different lakes with different Hg concentration in fish are compared and if these empirical data are normated for seasonal Hg fluctuations (only fish caught in a defined season are compared) and size variation, the pair-wise correlation coefficient is r = 0.9 (Table 4, Håkanson, 1984b).

From data given in Table 3, we can calculate how much of the variation in Hg content in fish that can be explained from the dose alone. The answer is 10%, i.e. $r^2 = 0.1$. This means that for this particular empirical data set, the dose will only explain a small part of the response, the F(Hg) value. If we add one lake sensitivity factor, the bioproduction, to the dose, how much more of the variation in Hg content in fish will be explained? The answer is about 26 %. If we also account for pH, the degree of explanation will increased to 74 % (0.84^2*100). This means that the given formula explains about 74 % of the variability in Hg content in fish. This is a very high figure since the maximum degree of explanation that may be expected is $0.9^2 = 0.81 = 81$ %.

	W_1	Environmental parameter W_2	W_3	Dose parameter D	$E = f(D, W_2)$	$E = f(D, W_3)$	$E = f(D, W_2, W_3)$	Empirical E-data
	Area (km²)	pH	BPI	Hg$_{se}$/R$_g$ (ng g⁻¹ ds)	$F(Hg) = f(Hg_{se}, pH)$ BPI = const = 3.83	$F(Hg) = f(Hg_{se}, BPI)$ pH = const = 6.35	$F(Hg)$ (mg kg⁻¹)	$F(Hg)_{emp}$ (mg kg⁻¹)
Large Lakes								
Värmlandssjön	3582	7.2	2.9	730	1.1	1.6	1.3	1.1
Dalbosjön	2066	7.1	3.1	160	0.4	0.55	0.5	0.7
Vättern	1856	7.6	2.9	80	0.2	0.35	0.25	0.45
Blacken (Mälaren)	91	7.3	5.0	260	0.55	0.55	0.45	0.5
River Kolbäcksån Lakes								
Bysjön	5.1	6.8	4.2	340	0.75	0.75	0.7	0.6
Övre Hillen	4.4	7.1	4.9	2040	1.7	1.7	1.4	1.2
Leran	2.8	6.9	5.2	650	1.1	0.95	0.85	0.7
Södra Barken	20.5	7.0	5.1	470	0.85	0.8	0.7	0.7
Stora Aspen	5.9	7.2	4.9	470	0.85	0.85	0.7	0.45
$F(Hg)_{emp}$ N								
Värmland Lakes								
1.2–1.8 8	1.35	5.1	3.4	300	1.1	0.85	1.2	1.5
1.0–1.2 9	0.38	5.5	3.8	280	0.9	0.7	0.9	1.1
0.8–1.0 7	1.36	5.9	3.5	220	0.70	0.65	0.75	0.9
0.6–0.8 15	9.38	5.9	3.8	220	0.70	0.6	0.70	0.7
0 4–0.6 11	1.14	6.0	3.4	160	0.55	0.55	0.55	0.5
Småland/Halland Lakes								
1.2–1.7 5	0.31	4.8	3.3	300	1.2	0.85	1.3	1.5
1.0–1.2 3	0.23	5.9	2.5	290	0.8	1.1	1.2	1.0
0.8–1.0 9	1.07	5.3	3.1	260	0.9	0.8	1.1	0.9
0.6–0.8 19	0.42	5.7	3.9	250	0.8	0.65	0.75	0.7
$N = 18$								
$\bar{x}$	425.0	6.35	3.83	416	0.85	0.83	0.85	0.84
s_x	1000.8	0.87	0.86	438	0.35	0.35	0.33	0.33
r	−0.03	−0.61	−0.30	0.31	0.62	0.51	0.86	1.00
Corr. sign.	Not	99%	73%	75%	99%	97.5%	99%	
Residual term (R)	99.9	62.8	91.0	90.4	62.4	74.0	26.0	

$\bar{x}$ = mean value, s_x = standard deviation, r = correlation coefficient and R = the residual term, which is defined as $100(1 - r^2)$. Modified after Håkanson (1980b).

Table 3: The relationship between the effect parameter, the Hg content in pike, the dose parameter, the Hg content in surficial sediment, and three environmental parameters – lake area, pH and bio-production index (BPI). From Håkanson (1984b).

Lake	Sample 1 Weight (kg)	$F(Hg)_{emp}$ (mg kg⁻¹ ws)	Sample 2 Weight (kg)	$F(Hg)_{emp}$ (mg kg⁻¹ ws)
Runn	2.48	1.75	2.35	2.30
Esköklubb	0.84	1.45	0.77	1.15
Mjörn	1.18	1.00	1.10	0.77
S Björkfjärden	1.30	0.85	1.24	0.41
Esköklubb	0.49	0.82	0.43	0.91
Bolmen	1.48	0.67	1.30	0.53
Mjörn	1.34	0.54	1.30	0.94
Älvkarleö	1.03	0.52	1.02	0.45
S Björkfjärden	0.72	0.48	0.78	0.42
Öresjön	0.73	0.42	0.78	0.70
Åsnen	0.58	0.37	0.63	0.36
Skikkisjön	0.27	0.31	0.25	0.40
Ljusne strömmar	0.89	0.27	0.84	0.16
Labbas	0.48	0.22	0.44	0.16
St Lulevattnet	1.71	0.21	1.57	1.57
Sörvikssjön	0.71	0.19	0.74	0.27
Storvindeln	0.42	0.18	0.42	0.13
Hjälmaren	1.27	0.18	1.03	0.16
Ladtjojaure	2.82	0.17	2.45	0.23
Roxen	0.88	0.17	0.84	0.23
Siljan	0.22	0.16	0.25	0.39
Hjälmaren	0.87	0.15	0.82	0.17
V Ringsjön	0.86	0.15	0.91	0.15
Vombsjön	1.00	0.11	1.04	0.096
Vombsjön	0.89	0.084	0.85	0.12
$\bar{x}$ ($N = 25$)		0.46		0.48
s_x		0.42		0.48

Correlation coefficient: $r = 0.90$
Residual term: $R = 100 (1 - r^2) = 18.2\%$

Table 4: Comparison between two sets of empirical data on the Hg content in muscle in 1 kg pike, F(Hg)emp, from 25 lakes from different geographical, limnological and contaminational environments in Sweden. These data have been corrected for seasonal variability (all fish were caught during October–November 1966) and size (only fish of equal weight are compared). From Håkanson (1984b).

In this context we will not discuss presuppositions and limitations
concerning the given formula and the results. Instead we will assume that
the results could reflect the major factors governing the Hg content in
fish. The main point here is that the dose will only account for a
minor part of the response and that it is very important to account for
the environmental sensitivity factors that regulate the "road from dose
to response".

In this context we will also very briefy mention antagonistic effects and
more specifically how zinc probably may act antagonistically towards
mercury (Håkanson and Uhrberg, 1981; Lindeström and Grahn, 1982).

Abnormally low Hg concentrations in fish have been found in several lakes
highly polluted with mercury and other metals. The mean concentration in
1 kg pike in such lakes lies in the range from 0.15 down to 0.05 mg/kg,
which is very low indeed since the "natural" Hg content in fish from
"unpolluted" lakes is seldom lower than about 0.2 (Håkanson 1984b)

Fig. 10 exemplifies results for Zn and Hg from :

(A) "uncontaminated" lakes from northern Sweden.
(B) "uncontaminated" lakes from middle Sweden.
(C) lakes receiving waste water from mines.

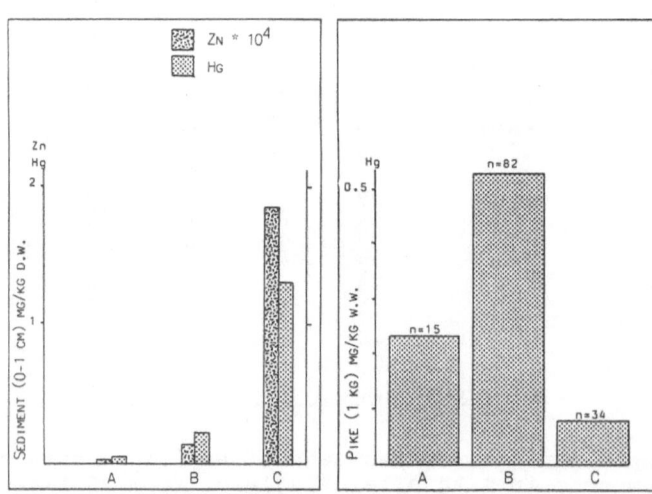

Figure 10: Zinc and mercury concentrations in surficial sediments and
methyl mercury in muscle in pike from (A) "uncontaminated"
northern Swedish lakes, (B) "uncontaminated" middle Swedish
lakes and (C) lakes receiving waste water from mines.

The sediment concentrations of several metals (Cu, Pb, Cd, Zn and Hg) are low in the lakes belonging to the (A)and (B) categories, but the Hg content in pike is much higher in these lakes than in the lakes belonging to the (C) category.

This is a most interesting situation where pollution of one comparatively harmless metal, Zn (or Se, see Björnberg et al., 1987), which is essential and does not biomagnify, may act antagonistically towards a very toxic metal, Hg.

REFERENCES

Andersson, T., Nilsson, Å., Håkanson, L. (1987). Kvicksilver i svenska sjöar (Mercury in Swedish lakes). Report SNV (preprint).

Beijer, K., Bengtsson, B-E., Jernelöv, A., Laveskog, A., Lithner, G., Westermark, T. (1977). Svenska vattenkvalitetskriterier - metaller. Del I, II (Swedish water quality criteria - metals). IVL Rapport B 398, Stockholm.

Björnberg, A., Håkanson, L., Lundbergh, K. (1987). A theory on the mechanisms regulating the bioavailability of mercury in natural waters. Manuscript. Dept. of Hydrology, Univ of Uppsala.

Chester, R. and Hughes, M. J. (1967). A chemical technique for the separation of ferromanganese minerals, carbonate minerals and adsorbed trace elements from pelagic sediments. Chem. Geol. 2

Förstner, U. and Wittman, G. T. W. (1979). Metal pollution in the aquatic environment. Springer - Verlag, Berlin.

Grahn, O. and Rosén, K. (1984). Deposition and transport of metals in some acid watersheds in southwestern, middle and north Sweden. (In Swedish). Swed. Env. Prot. Bd. SNV report 1687.

Grahn, O. and Sangfors, O. (1984). Temporal variation of heavy metals (Cd, Pb, Cu, Zn) in perch, zooplankton and water in four Swedish lakes with different metal loads and trophic status. (In Swedish). SERG report S 8401.

Håkanson, L. (1980). The quantitative impact of pH, bioproduction and Hg-contamination on the Hg-content of fish (pike). Env. Poll., 18.

Håkanson, L. and Uhrberg, R. (1981). Investigations in the River Kolbäcksån water system. XIII Metals in fish and sediments. (In Swedish). Swed. Env. Prot. Bd. SNV report 1408.

Håkanson, L. (1984a). Metals in fish and sediments from the River Kolbäcksån water system, Sweden. Arch. Hydrobiol. 101:3.

Håkanson, L. (1984b). Aquatic contamination and ecological risk - an attempt to a conceptual framework. Water res. vol. 18 no. 9.

Landner, L. and Grahn, O. (1975). Occurrence and behavior of heavy metals in three swedish rivers. (In swedish). Swe. Env. Prot. Bd. Final proj. report.

Lindeström, L. and Grahn, O. (1982). Antagonistic effects to mercury in some mine drainage areas. Ambio vol. 11 no. 6

Lindeström, L. (1984). Unpublished data. SERG.

Lindeström, L. and Qvarfort, U. (1985). Metals in soil water and surface water in the Garpenberg area - a study of the metal balance. SERG report S 8501.

Salomons, W. and Förstner, U. (1980). Trace metal analysis on polluted sediments. Part II: Evaluation of environmental impact. Environ Tech lett 1:506-517.

Salomons, W. and Förstner, U. (1984). Metals in the Hydrocycle. Springer Heidelberg 349 p.

Tessier, A., Campbell, P.G.C. and Bisson, M. (1979). Sequential extraction procedure for the speciation of particulate trace metals. Anal Chem 51:844-851.

EFFECTS OF pH ON THE UPTAKE OF COPPER AND CADMIUM
BY TUBIFICID WORMS (OLIGOCHAETA) IN TWO DIFFERENT
TYPES OF SEDIMENT

Anders Broberg and Gunilla Lindgren
Institute of Limnology, University of Uppsala
Box 557, S-751 22 Uppsala, Sweden

ABSTRACT

The uptake of Cu and Cd by tubificid worms (Oligochaeta) at pH 5, 6
and 7 was investigated using sediment from a low productive humic lake
and an eutrophic lake with high production.
The sediments were contaminated with Cu or Cd before the addition of
animals, which after incubation were analysed for Cu and Cd. Total
sediment, adsorbed/exchangeable fraction ($MgCl_2$), pore water and
overlaying water were analysed for Ca, Mn, Fe, Cu or Cd.
In both types of sediment a decrease in pH to 5 resulted in a markedly
increased concentration of the metals in all fractions. The
concentration of Cu or Cd in the labile fractions was very high with
sediment from the humic lake, which resulted in an almost momentanous
death of the added worms. The experiments with the eutrophic sediment
showed that a lower pH gave a smaller uptake of metals. The reasons
probably are decreasing surface adsorption, decreased uptake directly
through the skin and competition of binding sites. The differences in
uptake by the worms can also be due to pH-dependent variations in
supply of food and activity of the animals.

INTRODUCTION

The effect of heavy metals on aquatic organisms depends both on the
concentration of the metal and on its availability to the biota. In
the aquatic environment this availability is controlled by the
chemical form of the metal in water and sediment. Free metal ions and
ions adsorbed on particles are much more available to living organisms
than the fraction of the metal bound to organic material or to
different oxides (1, 2). Therefore, the content of, for example, Fe

and organics in the sediment can be very important for the availability of a metal (3).

The uptake of metals by the biota is also affected by factors such as pH. Often lower pH means increased availability and increased uptake of the metal (4, 5). These findings are contradicted by results of investigations showing that benthic fauna in acid lakes (pH 4.4) did not accumulate more metals than organisms in lakes with a pH-level one unit higher (6). The uptake of phosphorus by alga in experiments with Cd and Cu was stimulated by a lowering of pH (7) and a low pH also decreased the accumulation of Cd in mussels (8).

In this study the uptake of Cu and Cd by tubificid worms (Oligochaeta) was investigated in two different types of sediment and at pH 5, 6 and 7. The two sediments were quite different regarding nutrient level and contents of Fe, Mn and Ca. The aim was to see if there was any correlation between uptake of Cu or Cd, pH and concentration of the metal in two of the most available fractions of the sediment.

MATERIAL AND METHODS

Sediments from two lakes in Uppland, central Sweden, were used in the experiments. One of the lakes, Tarmlången, is a low productive humic lake whereas the other, Funbosjön, is a eutrophic lake with high production.

Contamination

The sediments were sieved (2 mm mesh size) and homogenized. The homogenized sediment was supplied with approximately 30 % tap water plus distilled water (50/50) and heated at 50 °C for 24 hours. The water was decanted and fresh water was added and the mixture stirred. The sediment was allowed to settle and once again the water was decanted. Each sediment was divided into three portions, of which one was shaken with 100 ml distilled water per l of sediment and the others with 100 ml distilled water containing 200 mg Cu ($CuSO_4.5\ H_2O$) or 100 mg Cd ($CdCl_2.2.5\ H_2O$). The different portions were placed in the dark at 20 °C and intermittently shaken for 5 days. After this period the portions were aerated for 17 days. During the aeration period pH was measured and adjusted in order to get the same pH in the different sediment portions.

Experimental procedure

The three sediment portions from each lake were further divided into 3
parts and pH was adjusted with sulphuric acid to 5, 6 and 7 both in
sediment and in lake water. For each lake and each pH experimental
series of 20 vessels, including 4 controls, was arranged. Each vessel
was supplied with 50 ml of sediment, 50 ml of lake water and 20-30
weighed tubificid worms. The worms came from an aquarium store and
consisted of 5-6 different species of tubificids. The experimental
series were incubated in the dark at 20°C and samples were taken 4
times during a period of 39 days. On each occasion worms from one
control and two vessels with Cu-contaminated sediment and two with
sediment containing Cd were collected in a sieve (0.2 mm mesh size).
The worms were starved for 24 hours in distilled water and their wet
and dry weights were determined. Sediment and overlaying water from
the last sampling were kept for analyses.

Analyses

Sediment and water (membrane filter 0.45 μm) were analysed for Fe,
Mn, Ca and Cu or Cd. The sediments were also analysed for water
content, carbon and nitrogen.

The pore water was separated from the sediment by centrifuging wet
sediment at 7500 rpm for 20 minutes. The supernatant was filtered
through a membrane filter (0.45 μm) and analysed for Ca, Fe, Mn and Cu
or Cd. In order to obtain the adsorbed / exchangeable fraction 2 g of
the residual after the centrifugation was extracted with 30 ml 1
mole/l $MgCl_2$, pH 7, for 1 hour. After centrifugation the liquid phase
was filtered through a 0.45 μm membrane filter and analysed for Ca,
Mn, Fe and Cu or Cd.

The worms were weighed on foil and dried at 105°C for 24 hours. The
dried animals were autoclaved for 30 minutes in 7 moles/l HNO_3. The
solution was diluted to 10 ml with distilled water, filtered through a
glassfiber filter (Whatman GF/C) and analysed for Cu or Cd.

The water content of the sediment was determined by drying overnight
at 105°C. C and N were obtained through combustion in a CHN-analysator
(Carlo-Erba) and the concentrations of the metals were measured with a
flame atomic absorption spectrophotometer Pye Unicam SP 192 or, at low
concentrations, with a graphite oven.

RESULTS

Sediment-chemical characteristics for the two lakes are shown in Table
1. The sediment from Lake Tarmlången was clearly humic with high water
content and large amounts of organic material. The content of N was,
however, unusually high for this type of sediment. The sediment in
Lake Funbosjön was largely affected by the production in the lake,
which means that the sediment can be characterized as gyttja. The
proportion of inorganic material was higher in the sediment from
Funbosjön than in sediment from Tarmlången, which can be seen from its
higher contents of Ca, Fe and Mn (Tables 1 and 2).

Table 1. Characteristics of Cu-contaminated sediments from Lake
Funbosjön and Lake Tarmlången.

	Funbosjön			Tarmlången		
	pH 5	pH 6	pH 7	pH 5	pH 6	pH 7
Water content (%)	87.4	85.9	86.8	96.5	96.6	96.5
Carbon (%)	7.66	7.72	7.86	22.1	21.7	22.0
Nitrogen (%)	0.80	0.76	0.76	1.88	1.83	1.88
Iron (mg/g dw)	45.0	44.3	46.3	28.5	31.5	31.8
Manganese "	0.61	0.68	0.70	0.24	0.28	0.30
Calcium "	3.3	3.3	3.7	0.6	0.8	1.0
Copper "	1.49	1.50	1.83	5.45	5.61	5.45

Table 2. Characteristics of Cd-contaminated sediments from
Lake Funbosjön and Lake Tarmlången.

	Funbosjön			Tarmlången		
	pH 5	pH 6	pH 7	pH 5	pH 6	pH 7
Water content (%)	87.0	87.2	86.6	96.7	96.6	96.5
Carbon (%)	7.66	7.72	7.86	22.0	22.6	22.2
Nitrogen (%)	0.80	0.76	0.76	1.87	1.89	1.86
Iron (mg/g dw)	42.5	42.3	50.2	28.8	31.3	33.9
Manganese "	0.61	0.69	0.80	0.24	0.26	0.36
Calcium "	3.5	3.5	3.9	0.6	0.8	1.3
Cadmium "	0.78	0.78	0.83	2.63	3.05	3.34

Tables 3 and 4 show the concentrations of Ca, Fe, Mn and Cu or Cd in
overlaying water, pore water and adsorbed/exchangeable fraction of the
sediment. In both types of sediment a decrease in pH to 5 resulted in
a markedly increased concentration of the metals in all 3 analysed
fractions. However, the concentration of Cu or Cd at pH 7 was higher
in the experiments with sediments from Lake Tarmlången than in
sediments from Lake Funbosjön at a pH of 5. On the other hand, the
water and pore water contents of Ca and Mn were considerably higher in
the experiments with sediment from Lake Funbosjön (Tables 3 and 4).

Table 3. Concentrations of certain metals in water, pore water and
         adsorbed/exchangeable sediment fraction in experiments with Cu.

		Funbosjön			Tarmlången		
		pH 5	pH 6	pH 7	pH 5	pH 6	pH 7
Water phase							
Cu	(ppb)	(1430)	150	75	9000	1350	445
Ca	(ppm)	266	183	62	54	44	8
Fe	(ppm)	0.09	0.03	0.02	0.06	0.06	0.06
Mn	(ppm)	17.6	2.0	0.01	2.2	0.9	0.04
Pore water							
Cu	(ppb)	485	115	50	7750	1800	360
	(μg/g dw)	2.1	0.5	0.3	81.5	18.3	7.2
Ca	(ppm)	327	510	–	59	30	–
Fe	(ppm)	1.0	0.36	–	0.36	1.8	–
Mn	(ppm)	20.5	7.3	–	2.6	1.2	–
Adsorbed/exchangeable fraction							
Cu	(μg/g dw)	120	50	45	1355	1195	480
Ca	(mg/g dw)	2.7	2.9	–	1.7	2.3	–
Fe	(μg/g dw)	20	10	–	20	5	–
Mn	(μg/g dw)	180	140	–	80	85	–

Worms added to contaminated sediment from Lake Tarmlången died immediately and therefore no results on metal uptake are available for this type of sediment. The experiments with sediment from Lake Funbosjön showed that the tubificid worms increased their content of metals with increased length of the incubation period (Figures 1 and 2), except for Cd at pH 5 . The uptake of Cu was larger at pH 7 than at lower pH and with Cd the uptake was lower at pH 5 than at higher pH

In some cases there were large variations between the duplicates, partly owing to sediment particles included in the autoclaved sample and partly to some uncertainty when weighing objects with very low weights. An abnormally high value for Cu from the serie at pH 5 (11 days) is not taken into consideration.

Table 4. Concentrations of certain metals in water, pore water and adsorbed/exchangeable sediment fraction in experiments with Cd.

		Funbosjön			Tarmlången		
		pH 5	pH 6	pH 7	pH 5	pH 6	pH 7
Water phase							
Cd	(ppb)	1390	760	70	24600	10400	3000
Ca	(ppm)	453	337	107	73	44	15
Fe	(ppm)	0.12	0.03	>0.02	0.06	0.06	0.06
Mn	(ppm)	20	6.9	0.01	2.3	1.4	0.43
Pore water							
Cd	(ppb)	1110	(30)	180	42000	16800	9240
	(µg/g dw)	4.8	(0.1)	0.8	645	240	170
Ca	(ppm)	675	505	337	102	7	27
Fe	(ppm)	9.3	0.10	0.6	0.01	0.01	0.36
Mn	(ppm)	28.3	10.8	12.8	3.6	1.7	1.2
Adsorbed/exchangeable fraction							
Cd	(µg/g dw)	205	70	50	1970	2165	1660
Ca	(mg/g dw)	3.6	4.2	5.4	1.0	1.1	1.6
Fe	(µg/g dw)	10	5	65	5	5	15
Mn	(µg/g dw)	205	215	100	90	90	80

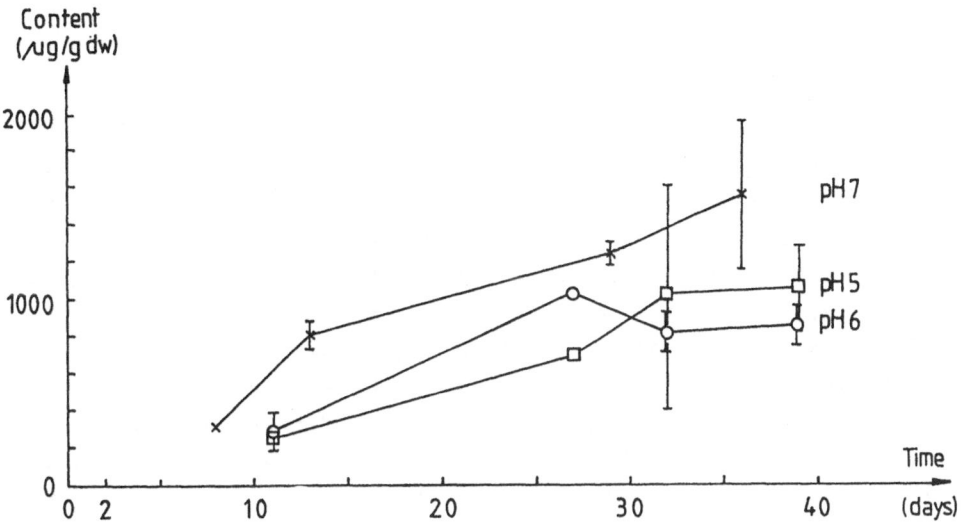

Figure 1. The effects of pH on the uptake of Cu by tubificid worms living in contaminated sediment from Lake Funbosjön.

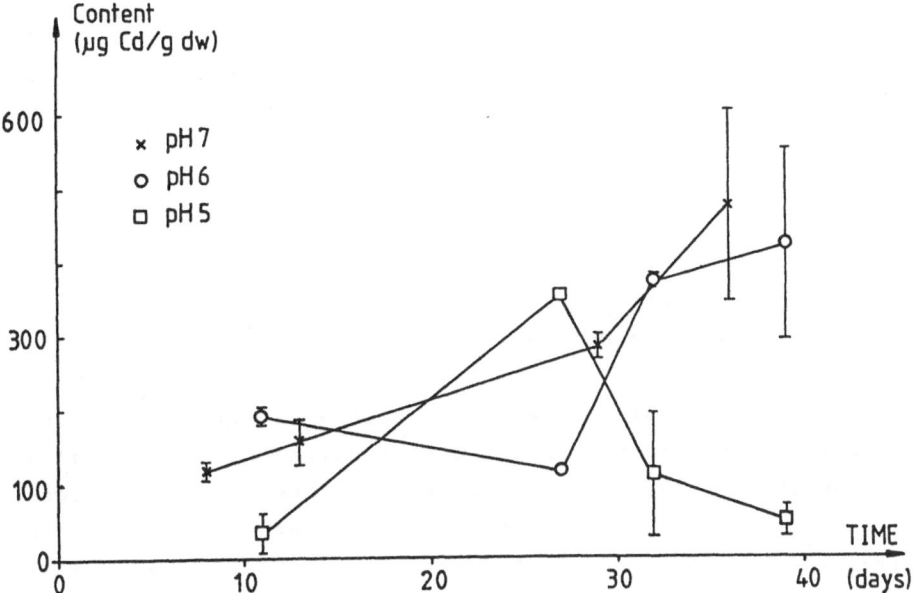

Figure 2. The effects of pH on the uptake of Cd by tubificid worms living in contaminated sediment from Lake Funbosjön.

DISCUSSION

The two sediments were quite different in character, dy in Lake
Tarmlången and gyttja in Lake Funbosjön (Tables 1 and 2). The metals
added were bound to the sediments to almost 100 % at the contamination
(pH 7). The high content of organics in sediment from Lake Tarmlången
should, according to, for example Tada & Suzuki (9) and Nriagu &
Coker (10), result in a binding of most of the metal added,
especially Cu, in relatively stable organic complexes with low
toxicity. In contrast, the proportions of Cu and Cd in pore water and
exchangeable fraction were clearly higher in Tarmlången than in
Funbosjön, 8 and 3 % of the total amount respectively for Cu (Table 3)
and 55 and 6 % for Cd (Table 4). One explanation might be that the
organic material in sediment from Lake Tarmlången has a relatively low
reactivity in the sense that the proportion of reactive groups, mainly
carboxyls and hydroxides, are small. Binding of the metals to the
inorganic part of the sediment, i.e., carbonates and hydroxides, may
be one reason for the lower concentration of the metals in the labile
fractions in sediment from Lake Funbosjön.

A lowering of pH in the contaminated sediments resulted in increased
concentrations of Cu and Cd in the two labile fractions and in
increased release of the metals to the overlaying water (Tables 3 and
4). This is quite in accordance with the results presented by, for
example, Sylva (11). The increased concentration of Cu in the labile
fractions can, according to Salomons & Förstner (12), be a result of
ion exchange:

$$CuX + H^+ \rightleftharpoons Cu^{2+} + HX$$

or of dissolution of substrate for binding, for example:

$$(Cu, Mn)O(OH) + H^+ \rightleftharpoons Mn^{2+} + Cu^{2+} + e^- + H_2O.$$

The contents of Fe in water were low (Tables 3 and 4), which probably
reflects relatively high concentrations of oxygen (not analysed) and
oxidized conditions. The lowering of pH to 5 gave a larger increase of
Cu (calculated as a percentage) in the pore water fractions, 7.5 and
11.7 times for Funbosjön and Tarmlången respectively, than in the
exchangeable fractions, 2.7 and 3 times respectively (Table 3).The
same was found, but on a less pronounced level, for Cd (Table 4).

In sediments from Lake Funbosjön the decreased pH not only gave higher
concentrations of Cu or Cd but also a very strong elevation of the

contents of Ca and above all Mn, which depended on an increased
solubility of carbonates and hydroxides (13). The contents of Ca, Fe
and Mn in sediment from Lake Tarmlången was considerably lower than in
sediment from Lake Funbosjön, implying that the lowering of pH gave
smaller increases of these ions in the pore water (Tables 3 and 4).

Relatively high concentrations of Ca and Mn were obtained in the
extraction with $MgCl_2$, which should give the fraction of the metal
loosely bound to the surface of the particles (14) (Tables 3 and 4).
This indicates that the extraction partly released Cu or Cd bound to
carbonates and easy reducible fractions, which are considered as being
more stable and less toxic than the adsorbed fraction of the metal.

The effect of a metal on aquatic organisms is often a function of the
concentration of free metal ion (15, 16). The concentration of Cu or
Cd in the labile fractions in the experiments with sediment from Lake
Tarmlången (Tables 3 and 4) was very high. Therefore the tubificid
worms died very soon after their addition to these sediments, whereas
they survived also at a pH of 5 in contaminated sediment from Lake
Funbosjön.

The uptakes of Cu and Cd by tubificid worms incubated in contaminated
sediment from Lake Funbosjön are shown in Figures 1 and 2. A lower pH
ought to be correlated to an increased accumulation of metals owing to
a higher degree of dissociation from organic and inorganic ligands and
increased availability of the metal to the organisms (17). This is
not in accordance with the results from Lake Funbosjön where, in
contrast, the uptake of Cu was largest at pH 7 and the worms
accumulated the highest amounts of Cd at pH 6 and 7. The reasons for
this discrepancy are probably the effects of pH on the different ways
in which a metal can be accumulated in organisms. Possible ways for
uptake are surface adsorption, uptake through integument and uptake
through ingestion (18, 19, 20).

The binding of metals to the surfaces of organic particles is
influenced by the supply of protonated sites on the membranes (21).
This implies that a lower pH results in a decreased uptake of metals
as surface adsorption and/or directly through the skin. Another reason
for the smaller accumulation of Cu or Cd at a lower pH is probably
competition of binding sites between the metals and other cations (
Ca, Mn), which were released from the sediment at the lowering of pH
(Tables 3 and 4). Such inverse relations between salinity and uptake

of metals are described for, e.g. macrophytes (22) and plankton
(23). In phase with the strongly increased concentrations of Ca and Mn
there was an increase in the contents of the corresponding anions,
which can partly, despite the low pH, bind metals in complexes and
thus depress the uptake. At the experiments with sediment from Lake
Tarmlången the concentration of competing ions was considerable less
than in Lake Funbosjön (Tables 3 and 4), which indicates a lower level
of site competition and complex binding and therefore a more
pronounced toxic effect of the metals.

The tubificid worms live in the sediment and feed on its content of
microorganisms and detritus (24), which means that the contents of
Cu or Cd in the sediment particles ought to be vital for the uptake of
metals. In the present case no large differences were found in the
total concentration of metals between contaminated sediments at
different pH (Tables 1 and 2), but the differences in uptake must be
due to variations in supply of food and activity of the animals, i.e.,
their rate of ingestion. As regards availability of food, the
bacterial biomass was considerably lower at a lower pH (Broberg,
unpubl.). Furthermore, at pH 7 the metal is bound to a greater extent
to carbonates and oxides at the sediment surface. These compounds
enter the intestine together with the food, are dissolved in the acid
environment and the metal can be taken up by the organisms (6). The
tubificid worms used in the experiments were taken from a neutral
environment. When the pH is lowered the animals are exposed to
physiological stress by which, among other things, they decrease their
ingestion of food. This stress was supposed to be the cause of a
smaller uptake of metals by mussels at a lower pH (8).

Other experiments with tubificid worms show that the uptake of Cu or
Cd by the organisms increases with the length of the incubation period
and is correlated to the concentration of metal in certain fractions
of the sediment (25). The change of pH in sediments from Llake
Funbosjön did not result in such correlations between certain metal
fractions and accumulation in the animals. The reason is that pH, as
described above, affected the patterns of metal uptake in the biota in
different ways. The largest uptake of metals at pH 7 (Figs. 1 and 2)
indicates that the ingestion of food was the most important source for
metals in the worms in this type of sediment. Most of the Cu or Cd
loosely sorbed to the surfaces of the organisms was released by
starving them in distilled water for 24 hours (26) and therefore
this amount was not included in the analysis of the total metal

content. Some authors (18, 19) consider the uptake of metals directly through the skin or by respiratory activity to be the most important routes of metal uptake in oligochaeta. At a pH of 7 these processes are not depressed by competition by other cations or protons, which happens at a lower pH. On the other hand, at pH 7 the concentration of metals in solution was considerably lower (Tables 3 and 4) and the degree of complexation higher, which is why these patterns for metal uptake are probably less important. Highest contents of metals in the organisms at pH 7 indicate the role of food ingestion in the accumulation of Cu and Cd by tubificid worms living in sediment from Lake Funbosjön. Furthermore, the importance of particulate metal for the accumulation in the worms will be quite clear by comparing Cu and Cd. The content of Cd in the labile fractions was higher than the concentration of Cu (Tables 3 and 4), whereas the total amounts of Cu in sediment and organisms (pH 7) were about twice the corresponding concentrations of Cd (Tables 1 and 2 and Figs. 1 and 2).

ACKNOWLEDGEMENTS

This work is a part of the project "Bioavailability and conversion of sediment-bound heavy metals" supported by the National Swedish Environment Protection Board.

REFERENCES

1. Förstner,U.,and G.T.W. Wittman. Metal pollution in the Aquatic Environment. (Berlin Heidelberg: Springer-Verlag, 2nd Ed,1981)

2. Jenne,E.A.,and S.N. Luoma. In: R.E. Wilding,and H. Drucker, Eds., Biological implications of metals in the environment. ERDA Symposium Series no 42, p 110-143 (1977).

3. Luoma,S.N.,and G.W. Bryan. J. Mar. Biol. Ass. U. K. 58:793-802 (1979).

4. Stokes,P.M.,and R.C. Bailey. In: Proc. Int. Conf. Heavy Metals in the Environment, Heidelberg, pp. 1049-1052 (1983).

5. Heit,M. In: Proc. Int. Conf. Heavy Metals in the Environment, Athens, pp. 655-657(1985).

6. Krantzberg,G. In: Proc.Int. Conf. Heavy Metals in the Environment, Athens, pp. 159-161(1985).

7. Peterson,H.G.,F.P. Healey,and R. Wagemann. Can. J. Fish. Aquat. Sci. $\underline{41}$:974-979 (1984).

8. Graney Jr.,R.L., D.S. Cherry,and J. Cairns Jr. Water Res. $\underline{18(7)}$:833-842 (1984).

9. Tada,F.,and S. Suzuki. Water Res. $\underline{16}$:1489-1494 (1982).

10. Nriagu,J.O.,and R.D. Coker. Environ. Sci. Technol.$\underline{14}$:443-446 (1980).

11. Sylva,R.N. Water Res. $\underline{10}$:789-792 (1976).

12. Salomons,W.,and U. Förstner. Metals in the Hydrocycle. (Berlin Heidelberg: Springer-Verlag, 1984).

13. Stumm,W.,and J.J. Morgan. Aquatic Chemistry .(New York: Wiley,2nd Ed, 1981).

14. Gibbs,R.J. Science $\underline{180}$:71-73 (1973).

15. Andrew,R.W., K.E. Biesinger,and G.E. Glass. Water Res. $\underline{11}$:309-315 (1977).

16. Sunda,W.G.,and P.A. Gillespie. J. Mar. Res. $\underline{37}$:761-777 (1979).

17. Allen,H.E.,R.H. Hall,and T.D. Brisbin. Environ. Sci. Technol. $\underline{14}$:441-442 (1980).

18. Back,H. In: Proc. Int. Conf. Heavy Metals in the Environment, Heidelberg, pp. 370-373 (1983).

19. Prosi,F. In: Proc. Int. Conf. Heavy Metals in the Environment, Heidelberg, pp.459-462 (1983).

20. Fowler,S.W. In: J. Salanki,Ed., Heavy metals in water organisms, Symposia Biologica Hungarica, $\underline{29}$:191-205 (1985).

21. Campbell,P.G.C.,and P.M. Stokes. Can. J. Fish Aquatic Sci. $\underline{42}$:2034-2049 (1985).

22. Rozema,J.,R. Otte,R. Broekman,and H. Punte. In: Proc. Int. Conf. Heavy Metals in the Environment, Athens, pp. 545-547 (1985).

23. Puddu,A.,M. Pettini,F. Bacciu,T. La Noce,and R. Pagnotta. In: Proc. Int. Conf. Heavy Metals in the Environment, Athens, pp.304-306(1985).

24. Pennak,R.W. Fresh-water invertebrates of the United States. (New York: Wiley,2nd Ed, 1978)

25. Diks,D.M.,and H.E. Allen. Bull. Environ. Contam. Toxicol. $\underline{30}$:37-43 (1983).

26. Taylor,D.,N.C.D. Craig,and D. Johnsson. In: Proc. Int. Conf. Heavy Metals in the Environment, Athens, pp. 652-655 (1985).

ALUMINIUM IMPACT ON FRESHWATER INVERTEBRATES AT LOW pH: A REVIEW

Jan Herrmann
Department of Animal Ecology, University of Lund
Ecology Building, S-223 62 LUND, Sweden

Abstract

The state of knowledge on aluminium (Al) impact on freshwater invertebrates at low pH is reviewed. Mainly inorganic ions seem to be biologically harmful. Published effect/mechanism descriptions may seem somewhat contradictory, but this can be due to the heterogeneity of "the invertebrate group", as well as the multitude and complexity of occurring Al species, thereby also Al analysis problems. Addition of Al to streams has in some cases increased drift and death of mainly some "surface-dependent" species (chironomids, mayflies, dance flies, dixid midges), but also some strictly benthic animals (isopods, stoneflies), while other studies on a variety of animals do not record any change in neither drift, mortality nor biomass.

In laboratory exposures Al has been shown to cause raised mortality for some daphnids and blackfly larvae at pH around 5; in the latter animals the effect was however mitigated by humus. A variety of other freshwater invertebrates were not affected. Moreover, at pH 4, Al has even been shown to improve the survival of mayfly nymphs and daphnids, otherwise impaired by the low pH in itself. The reason for this is not clear. Proofs for "food chain accumulation" of Al are still weak. Very high additions of Al have caused a decreased respiration rate in a dragonfly nymph, while a more field-relevant exposure level increased respiration in mayfly nymphs.This suggests a stress situation, probably due to impaired osmoregulation, indicating chemical or mechanical Al impact. The lowered oxygen uptake is then compensated for by improved respiration rate. A model for this is presented. Studies on crayfish, daphnids, mayflies and waterbugs indicate that Al can lower osmoregulatory efficiency and thereby affect the ion balance maintaining mechanisms of the animals. Such sublethal effects are important and should be studied further. The review also critically discusses the concepts mortality and bioavailability.

Introduction

Air pollution causes many environmental problems, and deposited acid substances are among the most important pollutants. In areas unable to neutralize acids, soil and surface waters show declining pH. There is an increasing number of acidified lakes and running watercourses in Europe and North America (1-3), and acidification has generally been accompanied by declining numbers of bottomliving invertebrates (4-7).

The pH levels of 4-5.5 also cause a rise of dissolved aluminium (Al) in the water, which seriously impaires most fish species, often with probably more serious effects just above pH 5 (8-10). Such observations have focused the attention of the scientists on that especially some insects, as mayfly nymphs and elmid beetles, but also some Crustaceans, seem affected in acidified waters. The reason is the idea that this possibly could be attributed the role of Al by increasing the susceptibility of benthic invertebrates to acidification (6, 11-12, 29-33, 35, 45). However, the co-occurence of high acidity and high levels of Al obscures the role of Al per se for the increased susceptibility of the animals. Also the mechanisms by which Al, under certain conditions, can act as a harmful agent are largely unknown.

Aluminium is one of the most common elements of the earth's crust (over 8%). Its chemistry and dynamics are complicated and far from sufficiently understood (e.g. 13). Al being insoluble as aluminium hydroxide and other forms around pH 6 but easily dissolved at pH of 4-5, a variety of Al ions dominate in acid natural waters. All these different forms - inorganic or organic, ions or solids, small or large molecules - are often called "species". It is generally accepted that of these the organic forms, i.e. when Al is complex bound to some large organic molecules, e.g. humus, are biologically relatively "harmless" (10, 14-16). Of the inorganic ions, Al^{3+} often dominate at low pH, but also unsaturated aluminium hydroxide, fluoride and sulfate complexes occur (15).

Unfortunately, in some studies dealing with the ecological effects of Al it is not clear whether total Al or some fraction (one or a group of species) thereof is meant by a denoted value. At least the total inorganic Al should be determined, preferentially limited to the monomeric fraction. Further, in some experimeltal studies the denoted

Al concentrations seem to refer to nominal values, i.e. what is
derived from the amount of Al added to the water, not the actually
dissolved concentration. This is unfortunate, as in many cases
probably a substantial part of the Al has precipitated (hydroxide at
pH around 5.5 - 6.0) or is complex bound to organic molecules. Both
these reactions, especially the latter, do probably lower the
"toxicity" of Al, which thereby probably becomes of less biological
importance. This can be overcome by measuring the Al concentration
throughout an experiment, and adding Al as compensation for these
physical processes. Otherwise, it could be recommended to report the
concentration value for the fraction of Al that is really dissolved.

In some acidified regions of Europe and North America, the Al content
in natural surface waters can reach up to 1.0 mg inorganic monomeric
Al L^{-1}, sometimes even more (12, 14, 17-20, J. Herrmann,
unpublished). Up to 10 mg Al L^{-1} has been reported for bog lakes, but
most of this is then probably complex bound to humus (21).

The aim of this paper is to present and discuss the state of
knowledge for the impact of Al at low pH on mainly benthic freshwater
invertebrates. The intention is then to put the emphasis on possible
and proposed effect mecahnisms, including connected problems, rather
than Al occurence descriptions. Havas (22) very recently reviewed Al
effects on all types of freshwater organisms, incl. algae and
vertebrates. Information on the effects of Al on fish can be found in
(8-9, 16, 22-27), while Ganrot (28) evaluates Al effects from medical
points of view.

Field studies

Field surveys of the occurence of benthic fauna and its relation to
abiotic factors do usually not give the most useful data on the
sensitivity of the animals to Al explicitly. This is because of the
strong correlation of Al to acidity (cf. above and 6, 11-12, 29-33),
and therefore less possibility to relate the animals occurence to
either acidity or Al. Some of these studies do however hint at the
possibility of Al may affect invertebrate populations in acid surface
waters, thus being partly responsible for declining species and/or
abundance numbers e.g. mayfly nymphs and elmid beetles. This way also
indicated for zooplankton at Al levels of 0.2 mg L^{-1} (34), whereas
field data for rotifers generally imply these animals to be rather

tolerant to Al (22).

However, "outdoor" experimentation in natural and artificial waters can give pertinent information. Thus, Hall and co-workers (35) simulated the conditions during the episodic Al increase during snowmelt by treating a second-order stream with Al (up to 3.8 mg L^{-1}; pH down to 4.8). They then noted an increased drift of invertebrates with increasing Al concentration. Especially animals related to the surface film, as the hyponeustic dixid midges, the emerging and egg-laying mayflies, as well as water surface feeding dance flies were found dead. It was suggested that Al caused a detrimental reduction of the surface tension, also indicated by a pronounced foam production. As the rise of the Al content concurrently also caused a substantial drop in pH, the interpretation of these results is in this case, however, uncertain as to the specific role of Al.

In an earlier study, Hall and Likens (36) lowered the stream pH, this also causing raised Al values and invertebrate drift, but the specific role of Al in this context is not clear. In other field experiments Hall et al. (37) showed that Al addition at pH 5-5.5 caused higher drift of mayfly nymphs, chironomids and dixid midges. The authors then concluded that these invertebrates are sensitive to Al at short-term pH depressions that by themeselves are supposed not to be stressful.

Ormerod and co-workers (38) treated two subsequent sections of a stream, one was brought to pH 4.3, and the next part to pH 5.0 plus an elevated Al content (0.35 mg inorganic Al L^{-1}). They also found that Al caused foam production. Very high drift respenses from the Al section was shown by the mayfly, Baetis rhodani, but also the drift values for Dixa and the mayfly Ephemerella increased significantly. Allard and Moreau (39), using artificial streams, noted increased mortality and drift of invertebrates due to an artificially lowered pH, from around 6.5 to 4.0. Adding also Al (0.4 mg L^{-1}) did however not cause any consistent additional rise in these parameters.

Excessive use of aluminium sulfate (alum) to precipitate phosphorus in water sewage treatment plants has also been shown to cause environmental hazards. One reason is that at pHs around neutrality high concentration of alum can result in dense floc layers of Al complexes on the bottom, and this can exert a physical stress on chironomids (40). Another reason is that the Al precipitations can be activated to aluminate ions, potentially toxic to fish, if pH rise from about 7 to 8-9 (41). Even if less relevant in the present"non-neutrality" context, this phenomenon should not be ignored.

Mortality experiments

A detrimental effect can exhibit as acute toxicity, chronic toxicity, or sublethal effects, though of course no distinct limits can be put between these concepts. Acute toxicity often implies mortality within a short time, often 96 h, while chronic toxicity appears after exposure during many days to weeks, thus for example preventing an animal to reach its reproductive stage. One possible delimination is to state that a chronic effect often continues for e.g. 1/10 of the organisms life span or more (G. Mackie, pers. comm.). In fact, in these expressions "mortality" should be used instead of "toxicity", as that is what is mostly meant by the latter word.Sublethal effects correspond to adverse impacts on processes as metabolism, excretion, respiration, osmoregulation, perception, movements etc. They cause an impaired condition, an unefficient resource utilization, or an increased energy consumption of the animals, thereby ultimately retarding growth and reproduction. This might be a more serious ecological impact by a pollutant on a population of some organism than if a considerable number of individuals are killed, as the latter might very well be compensated for (42).

Even if death is a drastic and simplified parameter of a substance impact, mortality experiments can give a lot of useful information of the susceptibility of different organisms. Thereby data can hint at appropriate studies of sublethal effects, towards which mechanism research always should aim.

That Al may cause immobilisation of Daphnia magna was reported already four decades ago, but without any pH information, the interpretation is uncertain (43-44). It has later, from recent acidification perspectives, frequently been suggested that the improverished invertebrate commmunities at pH around 4-5 could be due to the elevated Al levels (e.g. 35, 45). Thus it is expected that at a low pH, detrimental in itself, elevated levels of Al would, as for fish, cause additional mortality of benthic invertebrates. Certainly this has been shown to be valid in different experiments with Daphnia species, exposed to a variety of Al values from 1 (realistic) to 20 (very high level) mg Al L^{-1} (34, 45-46). Also, Ca has been shown to reduce Al toxicity to Daphnia (47), but only at pH 6.5, where almost all Al is in solid form.

Burton and Allan (48), using large outdoor artificial streams, report that Al = 0.5 mg Al L^{-1} in most cases cause additional mortality at pH 4.0 for the stonefly Nemoura sp. and the isopod Asellus intermedius, to some extent for the caddis larva Pycnopsyche guttifer, but not so for the caddis larva Lepidostoma liba and the snail Physella heterostropha. Survival was better at pH 5 and Al = 0.25 mg Al L^{-1}, as well as with high organic content of the water. No mayflies were tested. Further, levels of 0.2-0.5 mg Al L^{-1} seem lethal also to blackfly larvae at pH 4.8 (49). For the latter animals, raised humic content of the water counteracted the adverse effect of Al.

However, exposure to Al levels of 1 mg L^{-1}, and at various pH levels, did not cause any substantially increased mortality effects for the Crustacea Daphnia catawba, D. magna and Holopedium gibberum, as well as the insect larvae Chaoborus punctipennis and Chironomus anthracinus (47, 50). The exception was that at pH 6.5 the two Daphnia species in some experiments showed a significant reduction in vitality with increases Al levels. The authors then concluded that if these species decrease in numbers at low pH conditions, this is more probably due to biotic interspecific relationships than to acute H^{+} ion or Al toxicity (=mortality). Mackie (51) found that treatment of bivalves and gastropods with inorganic Al did not cause mortality at levels up to two orders of magnitude higher than found in most acidifying lakes of Ontario. Neither the amphipod Hyalella azteca, generally fairly sensitive, showed any consistently increaed mortality at high Al levels (G. Mackie, unpublished). However, as the author pointed out, sublethal detrimental effects can still be important. Three crayfish species, more or less affected by a pH just below 5, did not seem additionally influenced by an Al water content of 1-2 mg inorganic Al L^{-1} (52).

In laboratory experiments, carried out by myself, the survival of lotic mayfly nymphs were studied for 2-3 weeks in set-ups where Al as well as acidity were controlled and kept at different levels (53). The regimes were pH 4 and 5, combined with Al concentrations of 0, 0.5 and 2.0 mg inorganic Al L^{-1}.

As anticipated, in most experiments the mortality of the mayfly nymphs was significantly higher at pH 4 than it was at pH 5, thus not taking into account any Al impact (53). Heptagenia fuscogrisea seemed tolerant also at pH 4. However, at pH around 4 the presence of Al seemed to mitigate the detrimental impact of hydrogen ion stress. An example of one of these experiments is seen in Fig. 1. This ameliorating effect of Al was generelly more pronounced at 2.0 than at 0.5 mg Al L^{-1}. Further, the effect was less pronounced in early summer, close to the nymphs emerging period. The effect was clearly demonstrated for Heptagenia sulphurea, Ephemera danica and Leptophlebia marginata. Also the mayfly Baetis rhodani, very sensitive to low pH, responded in this way, whereas in the more acid-tolerant H. fuscogrisea the response was not clear. At pH around 5, the beneficial effect of Al was outlevelled or even reversed, thus more according to the expected effect.

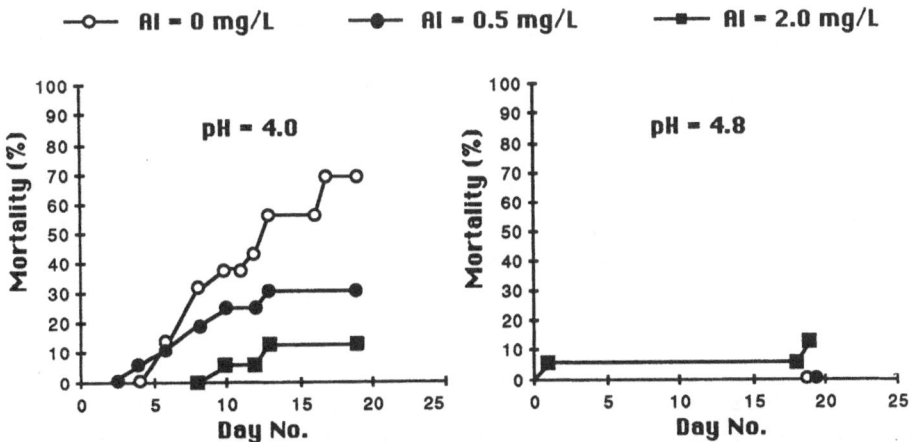

Figure 1: Mortality of nymphs of the mayfly Heptagenia sulphurea at different pH and Al conditions. Responses are significantly different (p ´ 0.02) for both pH and Al (for the latter at pH 4.0).

No published studies seem to exist, which demonstrate that insect survival at low pH, at least for some time, can be improved by Al. However, recently I was informed that also other mayfly nymphs can show this pattern (G. Raddum, unpublished; P. Greenwood, unpublished). Further, similar results, i.e. that Al can ameliorate the negative impact of low pH per se, have been reached with a few other test animals, as Daphnia magna (47, 54) and some fish and fish eggs (8-10, 55-57).

The mechanism behind this effect might be reduced membrane hydrogen ion permeability, changed sodium and/or calcium fluxes, or the buffering ability of Al (47, 54, 58). Another possibility was indicated by Correa et.al (59), who found that pH 4 caused the caddis larva Limnephilus sp. to increase its nitrogen (= protein) metabolism and loss, but also that this adverse effect was reduced by addition of 0.3 mg Al L^{-1}.

It has sometimes been put forward that Al like e.g. organochlorine residues and mercury may accumulate in animals, and that the rate of storage may be related to time and the animals level in the throphic web. No good proof of this does exist for the time being, but the phenomenon is presently dealt with (Herrmann and Andersson, in prep.). A "food chain transport" was suggested by Nyholm (60) to account for high levels of Al in the bone marrow and eggs of pied flycathchers. The birds were reported eating freshly emerged stoneflies from a nearby lake. A few analyses indicated that the insects could contain as much as 1.3 mg Al g^{-1} d w (E. Nyholm, unpublished). Even up to 4.9 mg Al L^{-1} have been reported for insects, but the validity of this value is not clear (cited in 22). Hall and Likens (36) provided a value of 0.84 mg Al g^{-1} f w for undefined "aquatic insects" in a H^{+}-treated small stream. Mayflies exposed in laboratory to 2.0 mg inorganic Al L^{-1} contained up to 0.6 mg Al L^{-1} d w (J. Herrmann, unpublished). Further, it can also be noted that benthic algae and macrophytes have been reported capable of accumulating Al (22, 61-62).

Ormerod and co-workers (63) found high stream Al content values where dippers, birds that mainly feed on caddis and mayfly larvae, breed less frequently. This might indicate a "true" food chain effect, provided that Al can act as a toxicant to the dippers, or an effect due to the absence of proper prey items for the birds, or just a co-variation due to other cause-effect relationships.

High levels of Al in insects from acid waters are not universal. Otto and Svensson (6) found the levels decreasing with increasing degree of maturity in a caddisfly. Further, the Al concentration was lower in animals kept under acid conditions that in those from slightly alkaline water, due to differences in solubility. The authors concluded that emerging caddis larvae could not substantially convey Al to predators, as leaving most Al behind in the exuviae.

Respiration studies

At low pH, high inorganic Al levels in natural waters cause Al hydroxide precipitation and overproduction of mucus - due to impaired osmoregulation - on the gills of many fish species (10, 64-65). This causes gill lesions, mechanical blockage for oxygen transport, leading first to hyperventilation, but later suffocation (8, 10, 25-26). All these effects seem most evident just above pH 5. The same mechanisms have therefore been expected to operate on some benthic invertebrates exposed to similar conditions.

However, very few studies exist on respiratory changes of invertebrates due to elevated acidity and Al impact. In recent experiments Herrman and Andersson (66) related the respiration rate to some kind of stress on the animals (cf. 9). The basic assumption was then that any adverse condition would increase respiration, to a limited extent, as was shown for naphtalene exposure to odonates (67). One must be aware, however, that mechanical blockage for oxygen transport could counteract and then "mask" this stress response.

Mayfly nymphs were exposed to 0, 0.5 and 2.0 mg Al L^{-1} at pH 4.0 and 4.8 for 10 days, and then the oxygen consumption rate of the nymphs was measured. The animals showed significant increase in weight-specific respiration with raised Al concentration (66). The mayfly species E. danica, which is restricted to the less heavily acidified region of South Sweden, was most affected by Al, as measured in this way. The responses of H. sulphurea and H. fuscogrisea to elevated Al was less pronounced, in accordance with these two species occurring also at more acid localities. Interesting enough, the effect of pH alone seemed less important for respiration. Exposures of caddis larvae (Limnephilus sp.) to ph 4.0 and pH 4.0$\pm$0.3 mg Al L^{-1} did in neither case affect respiration, compared to the level at pH 6.8 (59).

Further, Herrmann and Andersson (66) presented a tentative cause-effect relationship model for Al impact routes on mayfly nymphs (Fig. 2). Its essence is that mucus and/or hydroxides physically can prevent access to oxygen, hence lowering the uptake efficiency of oxygen over the epithelia (not only on gills, but most of the body). Compensation for this is attained by increased respiration rate (hyperventilation), causing stress. This response was termed the "mechanical impact route". Alternatively, the impaired osmoregulation, that is a less efficient ion transport over the membranes, together with other toxic effects form a "chemical impact route". Also here the impeded oxygen transport efficiency should be compensated for by an increased respiration rate. In both cases, the result would be increased energy demands or expenditures (cf. 68), resulting in less energy being available for growth and reproduction. Actually, the planktonic crustacean Daphnia magna showed in chronic tests for three weeks reproductively impairments of 16 and 50%, when exposed to 0.3 and 0.7 mg Al L^{-1}, respectively (46). The observations that Al leads to a higher respiration of mayflies (66), but with no change for caddisflies (59), fits with the idea that the increased respiration is a stress reaction due to affected osmoregulation, as caddis larvae are known to withstand also highly brackish waters with osmoregulatory problems.

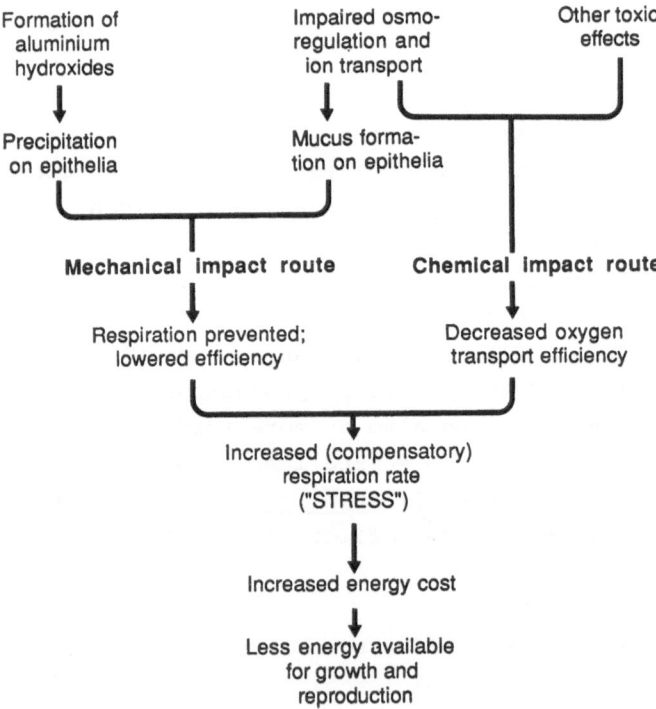

Figure 2: A proposed model for the possible physiological impact routes for Al effects on freshwater invertebrates at low pH conditions. From (61), with permission from D. Reidel Publishing Company.

Larvae of the dragonfly <u>Somatochlora</u> <u>cingulata</u> did however exhibit slightly decreased respiration rates with increased Al levels (69), but as the treatment levels were very high (10-30 mg Al L^{-1}), the interpretation of the data is uncertain. No other respiration study with relevant levels of Al seem to exist, but Diaz-Mayans and co-workers (70) found that increased exposure to lead, to some extent also cadmium, likewise caused a higher respiration rate of the crayfish <u>Procambarus</u> <u>clarkii</u>.

Thus, the way that Al might affect invertebrate respiration is far from clear. One reason for this might be that the direct effects of Al on small, mainly apneustic, animals are not comparable to those reported for fish. If respiration is a "mirror" of general stress related to various mechanisms, as suggested, it is rather an indirect and thereby more uncertain parameter of the influence of Al, as also suggested by Havas (68). One important direct effect, treated below, is the problem with maintaining the internal ion balance of the animal.

Osmoregulation studies

Cost and success for movements of ions - simple or complex, essential or "unwanted", passive or active - over a cell membrane are, among other things, largely dependent on and affected by pH conditions for many organisms. Generally, osmoregulation by freshwater organisms is considerably impaired by high acidity (54-55). Osmoregulatory inhibitions, their cause and effect, have been studies extensively on fish (8, 64, 71-72). It is now widely accepted that in many cases these problems are also due to the impact of Al (8-9, 23, 73).

Also for freshwater invertebrates in acid waters detrimental effects of increasing Al levels on osmoregulation was suspected by Otto and Svensson (6). There was however no indication of physiological stress in the caddis larva <u>Potamophylax</u> <u>cingulatus</u> under natural conditions, according to these authors. Exposure to realistic inorganic Al conditions (up to 1 mg Al L^{-1}) reduced the haemolymph Na^+ content in the crayfish <u>Astacus</u> <u>astacus</u> and <u>Pacifastacus</u> <u>leniusculus</u>, although with less adverse effects than those reported for fish (74). Inhibition of Ca^{++} uptake at pH 5.5 was regarded mainly due to the pH itself, only to a smaller extent affected by Al (up to 1 mg Al L^{-1}), thus causing an additional stress (55).

By using the ^{22}Na isotope, Havas and Likens (54) showed that Al affects both influx and outflux of sodium in D. magna. At pH 5 and higher there was a pronounced reduction of the influx but only small changes in outflux, so that the net result was an increased sodium loss. However, at pH 4.5 the influx did not change much when Al was added, whereas outflux was substantially reduced. This reduction of the net loss of sodium, although temporary, may be a reasonable explanation for the observed beneficial effects of Al, also observed by others (47, 53)(cf. the section on mortality).

The resulting effect of these fluxes (54) was found consistent with the observed Na contents. Thus addition of Al was reported to cause a Na decrease at pH 6.5, no change at pH 5.0, but an increase at pH 4.5. The authors (54) also reported an inverse relationship between Na content and mortality. Recently it was also found that Al can cause a decline of the Na content of mayfly nymphs at both pH 4.0 and 4.8 (76).

A pronounced reduction in Na influx at increased Al levels (up to 50 mg Al L^{-1}, a very high level) at pH 3 and 4 has also been demonstrated for the hemipteran waterbug Corixa punctata (21). A Na-flux from the tissue compartment to the haemolymph was suggested to compensate for this. The net result would be slightly increased Na levels of the haemolymph during the 20 h long experiment, and thus less harmful to the animal.

Potentially adverse effects of Al on osmoregulation is also supported by the observation that Al addition at low pH can cause increased Al-specific staining of anal papillae in phantom midges (77) and chironomids (M. Havas, unpublished). These structures are known to mediate the transport of cations, hence being vital to the salt balance processes. Al-indicating stainings have also been found in the anal region of mayflies, but the meaning is not clear (J. Herrman, unpublished).

Concluding remarks

It is perhaps not so strange that the results of the relatively few studies that exist regarding Al impact on freshwater invertebrates are somewhat contradictory, as the concentration of H^+ do largely affect the solubility and speciation of Al. Hence the patterns of occurence of the different forms become quite complicated, and the Al chemistry is still partly under debate (see e.g. 13, 24). Another reason is that the group of invertebrates of course is quite heterogeneous.

To detect detrimental effects of Al on freshwater invertebrates is generally not possible in field surveys, as a variety of potentially adverse factors are interrelated and changed concurrently. Nevertheless, field studies might hint at possible indirect Al effects on invertebrates, as changes in the physical habitat, food resources or predation pressure. Generally in all environmental impact research, it is important to stress that field studies of all types and laboraotry experiments "need" eachother, thus being complementary.

Mortality effects of Al on freshwater invertebrates do not seem to be as serious as anticipated from the knowledge on fish (cf. above). As concluded in the mortality section, sublethal effects and levels might actually be more important than those that kill some organisms. Nevertheless, there is probably a need of both types of approach.

A cautionary note could also be given on the word bioavailability. The reason is that this concept often implies the amount and type of a substance that potentially can be, or even really is, taken up by an organism. The substance might however be incorporated into the organism without any harm, or just pass throug the organism, all depending on the affinity, type and amount of the substance, but also type of organism, its feedings habits, developmental stage and exposure time. A measure of bioavailability, as attained from e.g. a dialysis bag, does therefore not necessarily say the whole truth about the biological potential or real activity of a substance.

As indicated earlier in this review, both mortality and respiration research suggest that studies of osmoregulatory effects on invertebrates are essential for the interpretation of impact of Al at low pH. Thus, in future studies of how increased Al levels in acidified surface freshwaters may affect these aquatic animals, a crucial process to study suggested to be osmoregulation; how it is affected, as well as how this affects the animals.

Acknowledgements

Valuable comments were kindly given by prof. P. Brink and Drs. K.G. Andersson and B.S. Svensson.

References

1. Drablos, D. and Tollan, A. Eds. Proc. Int. Conf. Ecol. Impact Acid Precip., 383 pp. (Oslo-Aas:SNSF-Project, 1980)

2. Haines, T.A. Trans. Am. Fish. Soc. 110: 669-707 (1981)

3. Wright, R.F. Water Quality Bulletin 8: 137-143 (1983)

4. Burton, T.M., Stanford, R.M. and Allan, J.W. In: F.M. D'Itri (Ed.) Acid Precipitation - Effects on Ecological Systems, pp. 209-235. (Ann Arbor: Ann Arbor Sci. Publ., 1983)

5. Singer, R. In: F.M. D'Itri, Ed. Acid Precipitation - Effects on Ecological Systems, pp. 329-363. (Ann Arbor: Ann Arbor Sci. Publ., 1983)

6. Otto, C. and Svensson, B.S. Arch Hydrobiol. 99: 15-36 (1983)

7. Okland, J. and Okland, K.A. Experientia 42, 471-486 (1986).

8. Muniz, I.P. and Leivestad, H. In: D. Drablos and A. Tollan, Eds., Proc. Int. Conf. Ecol. Impact Acid Precip., pp. 84-92 (Oslo-Aas: SNSF-project, 1980)

9. Muniz, I.P. and Leivestad. H. In: D. Drablos and A. Tollan, Eds., Proc. Int. Conf. Ecol. Impact Acid Precip., pp. 320-321 (Oslo-Aas: SNSF-Project, 1980)

10. Baker, J.P. and Schofield, C.L. Water, Air Soil Pollut. 18: 289-309 (1982)

11. Raddum, G.G. and Fjellheim, A. Verh. Internat. Verein. Limnol. 22: 1973-1980 (1984)

12. Simpson, K.W., Bode, R.W. and Colquhoun, J.R. Freshw. Biol. 15: 671-681 (1985)

13. Hooper, R.P. and Shoemaker, C.A. Science 229: 463-465 (1985)

14 Dickson, W. Verh. Internat. Verein. Limnol. 20: 851-856 (1978)

15. Driscoll, C.T., Baker, J.P., Bisogni Jr, J.J. and Schofield, C.L. Nature, 284: 161-164 (1980)

16. Brown, D.J.A. Bull. Environ. Contam. Toxicol. 30: 582-587 (1983)

17. Dickson, W. Vatten 39: 400-404 (1983)

18. Havas, M. In: M. Havas and J.F. Jaworski, Eds., Aluminium in the Canadian Environment, pp. 51-77. (Ottawa: National Research Council of Canada, Report No. 24759, 1986)

19. Nilsson, S.I. Ecol. Bull. 37: 120-132 (1985)

20. Lawrence, G.B., Fuller, R.D. and Driscoll, C.T. Biogeochemistry 2: 115-135 (1986)

21. Witters, H., Vangenechten, J.H.D., van Puymbroeck, S. and Vanderborgth, O.L.J. Bull. Environ. Contam. Toxicol. 32:575-579 (1984)

22. Havas, M. In: M. Havas and J.F. Jaworski, Eds., Aluminium in the Canadian Environment, pp. 79-127. (Ottawa: national Research Council of Canada, Report No. 24759, 1986)

23. Baker, J.P. In: T.A. Haines and Johnson, R.E. Eds., Acid Rain/Fisheries; Proc. Int. Symp. Acidic Rain and Fishery Impacts on Northeastern North America, pp. 165-176. (Bethesda: Am. Fish. Soc., 1982)

24. Odonell, A.R., Mance, G. and Norton, R. Water Research Centre Technical Report 197. (Medmenham, 1984)

25. Karlsson-Norrgren, L., Björklund, I., Ljungberg, O. and Runn, P. J. Fish Diseases 9: 11-25 (1986)

26. Karlsson-Norrgren, L., Dickson, W., Ljungberg, O. and Runn, P. J. Fish Diseases 9: 1-9 (1986)

27. Howells, G.D., Brown, D.J.A. and Sadler, K. J. Sci. Food Agricult. 34: 559-570 (1983)

28. Ganrot, P.O. Environ. Health Perspect. 65: 363-441 (1986)

29. Herrmann, R. and Baron, J. In: D. Drablos and A. Tollan, Eds., Proc. Int. Conf. Ecol. Impact Acid Precip., pp. 218-219 (Oslo-Aas: SNSF-Project, 1980)

30. Harriman, R. and Morrison, B.R.S. Hydrobiologia 88: 251-263 (1982)

31. Stoner, J.H., Gee, A.S. and Wade, K.R. Environ. Pollut. Ser. A. 35: 125-157 (1984)

32. Engblom, E. Lingdell, P-E. SNV pm 1798 (1985). (In Swedish with English summary)

33. Nilsson, A.N. and Johansson, A. Inform. Freshw. Res., Drottningholm 11/1985 (1985). (In Swedish with English summary)

34. Hörnström, E., Ekström, C. and Duraini, M.O. Rep. Inst. Freshw. Res., Drottningholm 61: 115-127 (1984)

35. Hall, R.J., Driscoll, C.T., Likens, G.E. and Pratt, J.M. Limnol. Oceanogr. 30: 212-220 (1985)

36. Hall, R.J. and Likens, G.E. Nature 292: 329-331 (1981)

37. Hall, R.J., Driscoll, C.T. and Likens, G.E. Freshw. Biol. 17 (1987, in press)

38. Ormerod, S.J., Boole, P. McCahon, C.P. Weatherley, N.S., Pascoe, D. and Edwards, R.W. Freshw. Biol. 17 (1987, in press)

39. Allard, M. and Moreau, G. Can. J. Fish. Aquat. Sci. 42: 1676-1680 (1985)

40. lamb, D.S. and Bailey, G.C. Bull. Environ. Contam. Toxicol. 27: 59-67 (1981)

41. Freeman, R.A. and Everhart, W.H. Tran. Amer Fish. Soc. 4:644-658 (1971)

42. Moriarty, F. Ecotoxicology. The Study of Pollutants in Ecosystems (London: Academic Press, 1983)

43. Anderson, B.G. Sewage Works Journal 16: 1156-1165 (1944)

44. Anderson, B.G. Trans. Amer. Fish. Soc. 78: 96-113 (1948)

45. Havas, M. and Hutchinson, T.C. Can. J. Fish. Aquat. Sci. 39: 890-903 (1982)

46. Beisinger, K.E. and Christensen, G.M. J. Fish. Res. Bd. Canada 29: 1691-1700 (1972)

47. Havas, M. Can. J. Fish. Aquat. Sci. 42: 1741-1748 (1985)

48. Burton, T.M. and Allan, J.W. Can. J. Fish. Aquat. Sci. 43: 1285-1289 (1986)

49. Petersen, R.C. Jr., Petersen, L.M.-M., Persson, U., Kullberg, A., Hargeby, A. and Paarlberg, A. Water Quality Bulletin 11: 44-49 (1986)

50. Havas, M. and Likens, G.E. Can. J. Zool. 63: 1114-1119 (1985)

51. Mackie, G.L. Amer. Malac. Bull. (1986, in press)

52. Berrill, M., Hollett, L. Margosian, A. and Hudson, J. Can. J. Zool. 63: 2586-2589 (1985)

53. Herrmann, J. (1987, manuscript to be submitted to Ambio)

54. Havas, M. and Likens, G.E. Proc. Natl. Acad. Sci. USA 82: 7345-7349 (1985)

55. Brown, D.J.A. J. Fish. Biol. 18: 31-40 (1981)

56. Dave, G. Ecotoxic. Environ. Safety 10: 253-267 (1985)

57. Neville, C.M. Can. J. Fish. Aq. Sci. 42: 2004-2019 (1985)

58. Johannesson, M. In: D. Drablos and A. Tollan, Eds., Proc. Int. Conf. Ecol. Impact Acid Precip., pp. 222-223 (Oslo-Aas: SNSF-project, 1980)

59. Correa, M., Coler, R., Yin, Y-M. and Kaufman, E. Hydrobiologia 140: 237-241 (1986)

60. Nyholm, E. Environ. Res. 26: 363-371 (1981)

61. Buergel, P.M. and Soltero, R.A. J. Freshw. Ecol. 2: 37-44 (1983)

62. Boylen, C.W. In: D.D. Adams & W.P. Page. Page, Eds., Acid Deposition. Environmental, Economic, and Policy issues, pp. 163-182. (New York, Plenum Press, 1985)

63. Ormerod, S.J., Allinson, N., Hudson, D. and Tyler, S.J. Freshw. Biol. 16: 501-507 (1986)

64. Fromm, P.O. Env. Biol. Fish. 5:79-93 (1980)

65. Grahn, O. In: Drablos and A. Tollan, Eds., Proc. Int. Conf. Ecol. Impact Acid Precip., pp. 310-311 (Oslo-Aas: SNSF-Project, 1980)

66. Herrmann, J. and Andersson, K.G. Water, Air Soil Pollut. 30: 703-709 (1986)

67. Correa, M. and Coler, R. Bull. Environ. Contam. Toxicol. 30: 269-276 (1983)

68. Havas, M. In R. Singer, Ed., Effects of Acidic Precipitation on Benthos, pp. 49-65. (Springfield, North American Benthological Society, 1981)

69. Correa, M., Coler, R.A. and Yin, C-M. Hydrobiologia 121: 151-156 (1985)

70. Diaz-Mayans, J., Torreblanca, A., Del Ramo, J. and Nünez, A. Bull. Environ. Contam. Toxicol. 36: 912-917 (1986)

71. Leivestad, H. and Muniz, I.P. Nature: 259: 391-392 (1976)

72. Rosseland, B.O. and Skogheim, O.K. Rep. Inst. Freshw. Res., Drottningholm 61: 186-194 (1984)

73. Baker, J.P. and Schofield, C.L. In: D. Drablos and A. Tollan. Eds., Proc. Int. Fonf. Ecol. Impact Acid Precip., pp. 292-293 (Oslo-Aas: SNSF-Project, 1980)

74. Appelberg, M. Hydrobiologia 121: 19-25 (1985)

75. Malley, D.F. and Chang, P.S.S. Arch. Environ. Contam. Toxicol. 14: 739-747 (1985)

76. Herrmann, J. Proc. Int. Symp. Ecophysiology of acid stress in aquatic organisms/Ann. Soc. Roy. Zool. Belg. (1987, in press)

77. Havas, M. Water, Air Soil Pollut. 30: 735-741 (1986)

Section 3

SUMMARY OF WORKING GROUP REPORTS

Lars Landner
Swedish Environmental Research Group
Götgatan 35, S-116 21 Stockholm

INTRODUCTION

During the second day of the "Workshop on the Speciation of Metals in Water, Sediment and Soil Systems", four separate Working Groups discussed different aspects of metal speciation in relation to different situations of metal discharge and metal contamination of the environment. Three Working Groups addressed the following three situations:

1) Metal speciation in relation to current wastewater discharges and atmospheric emissions from mining operations and metal processing (chairman: Professor Bert Allard, Linköping).

2) Metal speciation in relation to the dumping of solid waste, sewage sludge, dredging and other related activities (chairman: Professor Ulrich Förstner, Hamburg).

3) Metal speciation in relation to the mobilisation of metals from existing deposits in soil/sediments: acid rain, sulphide oxidation, etc (chairman: Professor Lars Håkansson, Uppsala).

The forth Working Group concentrated on analytical techniques and it was chaired by Dr. Brit Salbu, Oslo.

Among the various questions addressed by the three first mentioned Working Groups were for example:
- why do we need information on metal speciation ?
- what environmental factors control the distribution of species ?
- how can information on metal speciation be used to draw conclusions on the bioavailability and toxicity of metals ?
- what analytical procedures can be recommended ?
- what research needs can be identified ?

All four Working Groups prepared separate reports. However, since several overlaps occurred between the conclusions from the different groups, the editor has tried to make a synthesis of the major conclusions from all four Working Groups. It is hoped that the following summary will give a correct account of at least the most important points expressed during the Working Group discussions and during the following plenary session. If not, only the editor is to blame.

A complete list of participants of the Workshop is given in Appendix 1.

THE NEED FOR METAL SPECIATION

The release of metals from various metal handling activities (mining, metal processing, dumping of solid waste, etc) could lead to negative effects if these metals are distributed to, are taken up into, are accumulated by and are toxic to living organisms in the ecosystem. It is therefore essential for society to:
- evaluate and quantify potentially harmful effects of certain metals in the environment;
- predict future effects due to the presence and/or extended releases of metals from anthropogenic sources;
- design proper technical concepts for release control and deposition of metal wastes;
- understand basic processes in the ecosystem related to the distribution and mobilisation of metals;
- assess the chemical reactivity and transformations of critical species (bio-available and toxic species);
- identify species which are bioavailable (and possibly toxic) for various organisms.

A knowledge of metal speciation (including chemical speciation as well as physical characterisation) is required, since the speciation of the individual metal has an influence on:
- metal mobility - metal distribution in space and time;
- transport chains in the ecosystem - biological uptake, mineralisation, etc;
- chemical reactivity - effects of environmental parameters and changes in the environment on chemical processes;
- bioavailability and impact/toxicity for various organisms.

Since metal speciation is a laborious and expensive undertaking, it cannot be performed routinely as a part of all studies of metals in the environment. Therefore, we have to develop some simple priorities regarding when metal speciation should be undertaken.

An important parameter, to describe the major metal species in water, is the partition coefficient, K_p, for metal in solution, Me(s), and particle-bound metal, Me(p):

$$K_p = Me(s) / Me(p).$$

K_p is a function of pH, temperature, redox potential, alkalinity, salinity and content of humus, Ca, Fe, Mn and Al of the water. Most of these parameters are normally analysed within water quality monitoring programmes. Improved knowledge of how they influence K_p will aid in predicting major heavy metal speciation.

It is also obvious that metals are a quite heterogeneous group of elements with respect to their release, mobility and toxicity. Based on a set of simple criteria related to the occurrence, the natural concentrations, the release pattern, the mobility (solution and sorption behaviour) and the toxicity of the metals, the following priority setting regarding which metals should be speciated in the first place could be useful:

High priority group	Hg, Cd, As, Al
Medium priority group	Cr, Cu, Ni, Pb
Low priority group	Fe, Mn, Zn

FACTORS CONTROLLING THE DISTRIBUTION OF METAL SPECIES

The waste categories considered by the Working Groups were airbourne particles, wastewater and solid waste (ash, process sludge, tailings, high risk wastes) from mining operations and metal processing, dredged material, sewage sludge and household wastes.

Airbourne particles of two kinds would be expected:
- primary particles, frequently of low water solubility (e g high temperature fired oxides);
- secondary particles: metals associated with a carrier matrix.

Both types of particles will reach the terrestrial and the aquatic ecosystems, although the transport time and chemical state would be related to meteorological/ climatological factors as well as chemical factors (solubility and redox reactions in the air/precipitation system).

Liquid wastes would primarily be fairly acidic aqueous solutions with metals in ionic forms, metal sludges originating from the neutralization of acidic process solutions (e g mixtures of metal hydroxides and calcium sulphate) or deposits of precipitated dust and slag. The here mentioned solid wastes as well as dredged material, mine tailings, sewage sludge, etc are generally characterized by means of various tests for prediction of the long and short term fate of metals. Currently used characterisation methods or tests are: elutriate test, acid leachate test, chelator leachate test, cation exchange soil test, buffer capacity, base saturation test, phase characterisation (sequential leaching) before and after stabilisation and bioassays.

However, analysis and identification of metal species and of their bioavailability normally start when the wastes have reached the ecosystem, which would determine (or modify) the chemical state of the metals, in principle independently of the source.

The environment can be looked upon as a system of compartments in relation to metal speciation and metal transformation:
 * Aqueous solutions - metal species in true solution (ions, organic and in-

organic complexes).

* Non-stationary particulate and colloidal matter – true metal colloids (hydroxides) or pseudo-colloids (organic or inorganic carriers).
* Stationary solid systems – soil materials, sediments, etc.
* Organisms and detritus (dead biological material).

The speciation and distribution of a metal in and between these compartments are illustrated in Figure 1.

Chemical parameters that govern metal speciation are:

1. Redox potential (oxidation state)
2. pH (degree of hydrolysis, precipitation, sorption behaviour)
3. Complexing agents (OH^-, CO_3^{2-}, PO_4^{3-}, SO_4^{2-}, organics such as humic and fulvic acids)
4. Salinity, concentrations of competing elements
5. Temperature
6. Residence time

These factors are also significantly inter-related.

The sorption of species in solution would primarily depend on the degree of metal complexation (hydrolysis, complexes with inorganic or organic ligands) and properties of exposed solid surfaces (charge, presence of complexing groups, etc). Thus, also the distribution between solid surfaces and the solution phase can be described in terms of metal speciation (exchangeable metals, oxide-bound metals, organic-bound metals, co-precipitates, mineralised metals).

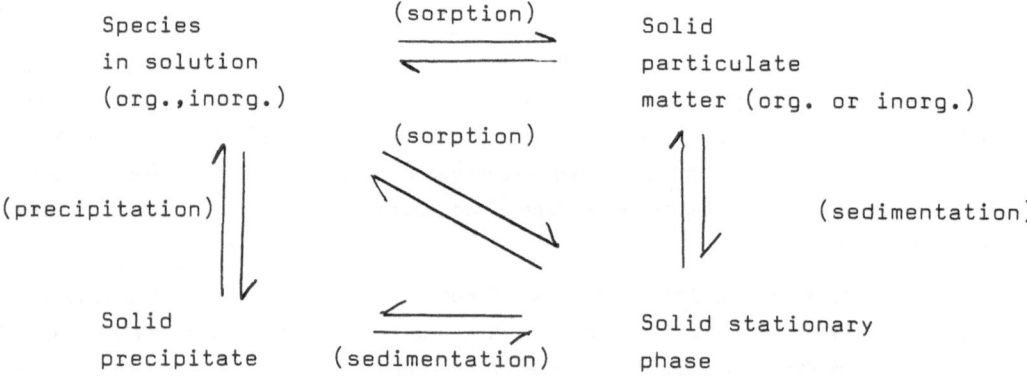

Figure 1. Distribution of species between solid and solution phases.

BIOAVAILABILITY AND TOXICITY

To be considered bioavailable a metal species must have a measurable uptake rate, resulting in bioaccumulation or a toxic response within the organism. Metal species presently recognized as bioavailable are low molecular weight species, mainly "free" and/or weakly bound ions but also small size molecules and complexes. The ultimate goal of chemical metal speciation is to identify and describe bio-available species (size, charge, ligands, etc), to identify the processes which transform other physico-chemical forms (reservoir - colloids, etc) into such species and to identify the factors (pH, redox potential, etc) that invoke such processes. It should also be held in mind that the bioavailability of a metal could be enhanced by the presence of organophilic ligands.

It is important to remember that bioavailability is a relative concept and that biological uptake and toxic impact of individual metals must be related to biological factors such as:
- type of organism
- developmental stage and sex
- activity patterns
- feeding strategy, etc.

More research is needed to establish good correlations between metal species and their bioavailability to different organisms.

ANALYTICAL TECHNIQUES

Routine analytical procedures - water

Most countries have routine procedures for water sampling and water analysis. In these standard procedures, metal analysis in general is based on either atomic absorption spectrometry (AAS) or inductively coupled plasma emission (ICP). In addition, some laboratories have standard procedures for filtration (0.45 um filter), centrifugation, or for the use of ion-exchange resins or extraction agents.

In order to obtain information on low molecular weight species, dialysis in situ is recommended as a standard routine technique. This dialysis technique re-duces the influence of colloids which affect results obtained, for example, from filtration, ion-exchange and extraction procedures. This standard procedure has to be specified in an informative manual. The laboratories using this procedure should also carry out inter-calibration tests.

Routine analytical procedures - sediments and soils

For routine analysis, the total metal concentration is usually analysed by strong acid digestion followed by atomic absorption determination.

In order to obtain information on the association of metals with different components present in soils/sediments, a three step sequential extraction scheme is recommended. The three fractions to be determined are:

1. Exchangeable fraction - based on extraction with magnesium chloride or ammonium acetate at pH 7.0.

2. Metals associated with the carbonate and hydrous metal oxide fraction - based on extraction with acidic hydroxylamine/hydrochloride.

3. Metals associated with organics and mineral lattice - based on digestion with strong acid/oxidation.

This scheme provides information on surface adsorbed metal species (fraction 1), the effect of reduced pH and redox potential (fraction 2), and the inert metal species (fraction 3).

Direct methods (analysis) are usually not informative with respect to metal species as the concentrations in natural waters are low and numerous compounds are present and may interfere with analytical signals. At present, combined techniques, i e different fractionation techniques interfaced with different detectors, where species are fractionated and then analysed with respect to total metal concentrations in different obtained fractions, are preferred for speciation purposes.

Procedures used in current research

A variety of techniques have been used and further developed at different research institutes. Metal species in water solutions can be fractionated according to, for example, size, charge, solubility, density, volatility and polarity. These different fractionation techniques can also be performed sequentially. These techniques are then interfaced with detectors giving information on the concentrations of metals and other elements (multielement methods) or information regarding structures (organic ligands) and isotopic composition. Along with the development of a useful combination, work is also conducted in order to simplify the techniques for routine analytical purposes.

RESEARCH NEEDS

It was concluded that basic research related to the speciation, distribution, biological uptake and effects of metals in the environment should be directed towards a basic understanding of the mechanisms behind the observed phenomena.

Important research needs can be recognised in the following areas:

1. Analytical methods for distinguishing various metal species:
   - methods fractionating species according to size and charge simultaneously
   - specific methods for lipid soluble metal species
   - methods to get information on the influence of colloids on metal species

2. Determination of the bioavailability:
   - for various elements/species
   - for various organisms
   - under different conditions
   - general properties of bioavailable metal species (charge, size, complexation)

3. The role of natural organics in metal speciation, transport and biological uptake:
   - properties of humic substances (e g functional groups)
   - synergistic effects (increased bioavailability of toxic metals)
   - increased mobility - mobilising agents

4. Design of model systems that in a controlled way can simulate environmental systems and be used for the study of biological uptake
   - choice of conditions
   - choice of organisms
   - use of fractionated natural waters in model ecosystem experiments

5. Development of useful standardised biomonitoring methods and assessment of their limitations

6. Mathematical modelling to obtain a better understanding of sorption/desorption phenomena:
   - quantitative modelling
   - coupling of sorption and chemical speciation

7. Modelling of combined transport and chemical/biological reactions of dissolved metals:
   - retardation of metals in various environments
   - fixation/mineralisation of metals

8. Development of cheap and effective technology for metal demobilisation to be used in solid waste disposal

9. Preparation of an informative manual describing a standardized procedure for metal speciation in natural waters.

Appendix

LIST OF THE PARTICIPANTS OF THE "INTERNATIONAL WORKSHOP ON SPECIATION OF METALS
IN WATER, SEDIMENT AND SOIL SYSTEMS, HELD AT SUNNE, SWEDEN, SEPTEMBER 15-16, 1986

ALLARD BERT
Department of Water, University of Linköping, S-581 83 LINKÖPING, Sweden

AZCUE JOSE M.P.
Laboratorio de Radioisotopas, Instituto de Biofisica, Universidade Federal
de Rio de Janeiro, Ilha do Fundao, RIO DE JANEIRO, Brazil

BJÖRNSTAD HELGE E.O.
Department of Chemistry, University of Oslo, P.O.B. 1033, Blindern,
N-0315 OSLO 3, Norway

BORG HANS
The National Environmental Protection Board, Trace Metal Laboratory, Box 1302
S-171 25 SOLNA, Sweden

BROBERG ANDERS
Institute of Limnology, University of Uppsala, Box 557,
S-751 22 UPPSALA, Sweden

FÖRSTNER ULRICH
Arbeitsbereich Umweltschutztechnik, Technical University of Hamburg-Harburg,
P.O. 90 14 03, D-2100 HAMBURG 90, Federal Republic of Germany

GRAHN OLLE
Swedish Environmental Research Group (SERG), Fryksta, S-665 00 KIL, Sweden

GRANDE MAGNE
Norwegian Institute for Water Research (NIVA), P.O. Box 333, Blindern,
N-0314 OSLO 3, Norway

GÖRANSSON TORBJÖRN
Boliden Mineral AB, S-936 00 BOLIDEN, Sweden

HERRMANN JAN
Department of Animal Ecology, University of Lund, Ecology Building,
S-223 62 LUND, Sweden

HULTBERG HANS
Swedish Institute for Water and Air Pollution Research (IVL), Box 5207,
S-402 24 GÖTEBORG, Sweden

HÅKANSSON LARS
Department of Hydrology, University of Uppsala, V. Ågatan 24,
S-752 20 UPPSALA, Sweden

IVERSEN EIGIL
Norwegian Institute for Water Research (NIVA), P.O. Box 333, Blindern,
N-0314 OSLO 3, Norway

LANDNER LARS
Swedish Environmental Research Group (SERG), Götgatan 35,
S-116 21 STOCKHOLM, Sweden

LEHTINEN KARL-JOHAN
Finnish Environmental Research Group (FERG), Torpvägen 13-15 B7,
SF-01150 SÖDERKULLA, Finland

LINDESTRÖM LENNART
Swedish Environmental Research Group (SERG), Fryksta, S-665 00 KIL, Sweden

LYDERSEN E.
Department of Chemistry, University of Oslo, P.O. Box 1033, Blindern,
N-0315 OSLO 3, Norway

MORRISON GREG
Department of Sanitary Engineering, Chalmers University of Technology,
S-412 96 GÖTEBORG, Sweden

PÄRT PETER
Department of Zoophysiology, University of Uppsala, Box 560,
S-751 22 UPPSALA, Sweden

RASMUSON ULRIKA
Technical Department, The National Environmental Protection Board, Box 1302,
S-171 25 SOLNA, Sweden

REUTHER RUDOLF
Swedish Environmental Research Group (SERG), Fryksta, S-665 00 KIL, Sweden

ROSEMARIN ARNO
AMBIO, Royal Swedish Academy of Sciences, Box 50005, S-104 05 STOCKHOLM, Sweden

SALBU BRIT
Department of Chemistry, University of Oslo, P.O. Box 1033, Blindern,
N-0315 OSLO 3, Norway

SANGFORS OLOF
Swedish Environmental Research Group (SERG), Fryksta, S-665 00 KIL, Sweden

SVEDBERG ROLF
Boliden Metall AB, S-932 00 SKELLEFTEHAMN; Sweden

TIMM BIRGITTA
Department of Research and Development, The National Environmental Protection
Board, Box 1302, S-171 25 SOLNA, Sweden

WALTERSON EVA
Swedish Environmental Research Group (SERG), Götgatan 35,
S-116 21 STOCKHOLM, Sweden

WESTLING OLLE
Swedish Institute for Water and Air Pollution Research (IVL), Aneboda,
S-360 30 LAMMHULT, Sweden

ÅNES KARL JAN
Norwegian Institute for Water Research (NIVA), P.O. Box 333, Blindern,
N-0314 OSLO 3, Norway.